TRANSPORTATION ISSUES, POLICIES AND R&D

CONGESTION IMPACTS OF FREIGHT FLOWS

TRANSPORTATION ISSUES, POLICIES AND R&D

Additional books in this series can be found on Nova's website under the Series tab.

Additional e-books in this series can be found on Nova's website under the eBooks tab.

TRANSPORTATION ISSUES, POLICIES AND R&D

CONGESTION IMPACTS OF FREIGHT FLOWS

DIANA HUGHES
EDITOR

NOVINKA
New York

Library of Congress Cataloging-in-Publication Data

ISBN: 978-1-53610-033-4

Published by Nova Science Publishers, Inc. † New York

CONTENTS

Preface		**vii**
Chapter 1	U.S. Border Communities: Ongoing DOT Efforts Could Help Address Impacts of International Freight Rail *United States Government Accountability Office*	**1**
Chapter 2	Freight Transportation: Developing National Strategy Would Benefit from Added Focus on Community Congestion Impacts *United States Government Accountability Office*	**35**
Index		**99**

PREFACE

About 93 trains a day on average crossed into the continental United States from Canada and Mexico in 2014, according to U.S. Department of Transportation's (DOT) Bureau of Transportation Statistics (BTS). Trains enter and leave the United States through 30 port of entries (POEs)—23 on the northern border and 7 on the southern border. Although international freight rail plays an important role in U.S. economic and trade interests, the movement of rail through U.S. communities at the border can result in blocked highway-rail grade crossings and vehicle traffic congestion. This book describes the factors that affect the movement of freight rail and the actions taken by federal agencies and others to expedite freight rail in selected POEs; and examines what is known about the impacts of freight rail operations on highway-rail grade crossings in POE communities. Moreover, the book addresses recent changes in U.S. rail and truck freight flows and the extent to which related traffic congestion is reported to impact communities; and the extent to which DOT's efforts to implement Moving Ahead for Progress in the 21st Century Act (MAP-21) address freight-related traffic congestion in communities.

In: Congestion Impacts of Freight Flows
Editor: Diana Hughes
ISBN: 978-1-53610-033-4

Chapter 1

U.S. BORDER COMMUNITIES: ONGOING DOT EFFORTS COULD HELP ADDRESS IMPACTS OF INTERNATIONAL FREIGHT RAIL*

United States Government Accountability Office

WHY GAO DID THIS STUDY

About 93 trains a day on average crossed into the continental United States from Canada and Mexico in 2014, according to DOT's Bureau of Transportation Statistics (BTS). Trains enter and leave the United States through 30 POEs—23 on the northern border and 7 on the southern border. Although international freight rail plays an important role in U.S. economic and trade interests, the movement of rail through U.S. communities at the border can result in blocked highway-rail grade crossings and vehicle traffic congestion. House Report 113-464 accompanying the Departments of Transportation, and Housing and Urban Development Appropriations Act included a provision for GAO to review the impact of international rail crossings on U.S. border communities.

This report (1) describes the factors that affect the movement of freight rail and the actions taken by federal agencies and others to expedite freight rail

* This is an edited, reformatted and augmented version of a United States Government Accountability Office, Publication No. GAO-16-274, dated January 2016.

in selected POEs and (2) examines what is known about the impacts of freight rail operations on highway-rail grade crossings in POE communities. GAO visited four POE communities that were selected in part based on BTS's 2010–2014 data on average incoming train volume. In each POE, GAO interviewed officials from local and state governments, the railroad, CBP, and FRA. GAO also interviewed officials from DOT, CBP, the Border Trade Alliance, and the Association of American Railroads.

WHAT GAO RECOMMENDS

GAO is not making recommendations in this report. DOT and CBP provided technical comments, which were incorporated.

WHAT GAO FOUND

Factors such as inspections and crew changes affect freight rail movements in the four U.S. border port of entry (POE) communities GAO visited, which can result in blocked highway-rail grade crossings. Federal agencies and others have taken actions to expedite rail in these communities. As part of its mission to safeguard the border, U.S. Customs and Border Protection (CBP) scans inbound rail cars on both borders using the Rail Vehicle and Cargo Inspection System (RVACIS), a machine used to detect anomalies and threats to national security. CBP generally requires trains to slow in order to pass through R-VACIS. To expedite freight rail and reduce blocked highway-rail grade crossings, CBP, for example, adjusted its procedures to allow certain trains to go through R-VACIS faster at two POEs on the northern border. Similarly, crew changes can result in stopped trains and blocked U.S. highway-rail grade crossings, particularly on the southern border. U.S. Department of Transportation (DOT) officials stated that crew changes are required due to differences in safety regulations between the U.S. Federal Railroad Administration (FRA) and Mexico. Railroads have expressed interest in eliminating such crew changes but face challenges such as FRA and labor union safety concerns.

The impacts of international freight rail on highway-rail grade crossings in communities GAO visited vary based on border-specific factors and community characteristics, and DOT improvement efforts including the issuance of a final rule could provide better data for help determining these impacts in the future. Inspections and crew changes, as well as rail traffic levels, can vary across POEs. For example, some factors play a role at southern, but not northern POEs. In addition, freight rail impacts vary based on community characteristics such as the availability of overpasses. State and local officials face data limitations, which reduce their ability to quantify rail-related community impacts. For example, local officials often do not have data on the number and length of trains passing through the community. In September 2014, GAO recommended that DOT improve the availability of national data to assess freight impacts on traffic congestion. DOT agreed and has actions under way. In January 2015, the FRA issued a final rule requiring railroads to update FRA's highway-rail crossing inventory once every 3 years. Prior to this rule, railroads voluntarily submitted data that were not always updated. DOT data efforts could better equip state and local governments to define the extent of blocked highway-rail grade crossings in communities nationwide, including at rail border communities.

ABBREVIATIONS

AAR	Association of American Railroads
BLET	Brotherhood of Locomotive Engineers and Trainmen
BTS	Bureau of Transportation Statistics
CBP	Customs and Border Protection
DOT	Department of Transportation
FRA	Federal Railroad Administration
MAP-21	Moving Ahead for Progress in the 21st Century Act
MPO	metropolitan planning organization
POE	port of entry
R-VACIS	Rail Vehicle and Cargo Inspection System

* * *

January 28, 2016

The Honorable Susan Collins
Chairman
The Honorable Jack Reed
Ranking Member

Subcommittee on Transportation, Housing
and Urban Development, and Related Agencies
Committee on Appropriations
United States Senate

The Honorable Mario Diaz-Balart
Chairman
The Honorable David Price
Ranking Member
Subcommittee on Transportation, Housing
and Urban Development, and Related Agencies
Committee on Appropriations
United States House of Representatives

Approximately 34,000 trains—about 93 trains a day on average—crossed into the continental United States from Canada and Mexico through 30 ports of entry (POE) in 2014, according to the U.S. Department of Transportation's (DOT) Bureau of Transportation Statistics (BTS).[1] The vast majority of these trains carry freight such as chemicals, lumber, and manufactured goods.[2] According to BTS, freight rail carried about 15 percent of the total value of all U.S. freight flows between the United States and Canada and Mexico in 2014. In that year, trucks carried the majority (about 60) percent of these freight flows, which amounted to $1.2 trillion worth of freight, in total.

Although international freight rail plays an important role in U.S. economic and trade interests, the movement of rail through U.S. border communities where POEs are located can temporarily block highway-rail grade crossings and contribute to traffic congestion. We have previously reported that overall freight rail traffic has increased since 2009 and may exacerbate traffic congestion concerns in many communities nationwide.[3] In addition, due to customs inspections and other processes at rail POEs, communities in these areas may face additional time that highway-rail grade crossings are blocked. In particular, as trains enter the United States, they are subject to inspections by the Department of Homeland Security's U.S. Customs and Border Protection (CBP). Trains entering from Mexico are also subject to equipment safety inspections required by the Federal Railroad Administration (FRA). Similarly, freight trains leaving the United States may be subject to inspections by Canadian or Mexican customs agencies. In addition, crew changes may occur, when the train is handed off between foreign and U.S. crews. As a result, trains may travel at slow speeds through or temporarily stop in rail POE communities. When this occurs as trains travel through highway-rail grade crossings, vehicle traffic must wait for the train to clear, potentially resulting in queues of vehicles, wait times, and increased congestion.

The House Report accompanying the Departments of Transportation, and Housing and Urban Development, and Related Agencies Appropriations Act of 2015 included a provision for us to review international rail border crossing times and the blockage of highway-rail grade crossings on the U.S. side.[4] This report (1) describes the factors that affect the movement of freight rail through selected ports of entry and the actions taken by federal agencies and others to expedite freight rail in these locations, and (2) examines what is known about the impacts of freight rail operations on highway-rail grade crossings in U.S. port of entry communities.

To determine the factors that affect the movement of freight rail and the impacts of freight rail operations on highway-rail grade crossings in U.S. border communities, we selected nine rail POE communities—Nogales, Arizona; El Paso, Eagle Pass, Brownsville, and Laredo, Texas; Blainc, Washington; Ranier, Minnesota; Port Huron, Michigan; and Rouses Point, New York. We selected communities that had at least one incoming train per day from 2010 through 2014 based on BTS border crossing data and excluded certain rail POEs, such as those outside of the continental United States or those with largely grade-separated infrastructure, meaning the rail line rarely intersects with vehicular traffic.[5] Of these, we conducted visits to four rail POE communities—Brownsville and Laredo, Texas; Ranier, Minnesota; and Blaine, Washington—that were selected to include communities with heavy inbound train traffic and a mix of northern and southern border locations. At each site visit, we interviewed representatives from the city or county, the metropolitan planning organization (MPO, if applicable),[6] the state department of transportation, CBP, FRA regional office, and the Brotherhood of Locomotive Engineers and Trainmen (BLET)—a union representing train operators. We also interviewed representatives from the five railroads that operate trains passing through each of the four rail POE communities we visited. For the remaining five of nine selected communities that we did not visit, we interviewed local officials by phone.[7] Furthermore, we interviewed officials and reviewed documents from CBP, DOT, FRA, and Department of State and interviewed representatives of the American Association of State Highway and Transportation Officials, the Border Trade Alliance, and the Association of American Railroads (AAR). To examine what is known about the impacts of international freight rail operations on highway-rail grade crossings, we reviewed relevant DOT documentation such as the reporting requirements for the National Highway-Rail Crossing Inventory and interviewed DOT officials on available data sources. To estimate the total time highway-rail grade crossings are blocked in eight of the nine selected rail POE communities;[8] we calculated the average time that freight trains would block key intersections in these communities based on the average speed of trains, length of trains, and frequency of trains that were reported by railroad representatives. We attempted to collect data from five railroads,[9] but we received incomplete information in response and were able to analyze information from two of these railroads.[10] Finally, we observed the CBP inspection process and the geography and relevant highway-rail crossings in each community we visited to gain additional insights related to international freight rail and the related POEs.

We conducted this performance audit from February 2015 to January 2016 in accordance with generally accepted government auditing standards. Those standards require that we plan and perform the audit to obtain sufficient, appropriate evidence to provide a reasonable basis for our findings and conclusions based on our audit objectives. We believe that the evidence obtained provides a reasonable basis for our findings and conclusions based on our audit objectives.

BACKGROUND

Canada and Mexico are the United States' first and third largest trading partners, respectively, and most freight between the United States and these countries is transported by truck and rail. Freight trains include bulk freight and intermodal freight. Bulk freight—such as grain, automobiles and component parts, coal, and chemicals—are transported in rail cars. For example, railroads deliver automotive parts made in the United States to assembly plants in Mexico by rail, and return finished automobiles from Mexico by rail. In addition, according to AAR representatives, bulk freight such as grain and lumber enters the United States along the northwestern border with Canada. Further, "intermodal" freight consists of containers carried by rail and transferred to or from other transportation modes, such as ships or trucks. For example, intermodal freight containers arrive at Prince Rupert in western Canada from Asia by ship and are transferred to rail and exported to the United States, entering through Ranier, Minnesota. Intermodal freight generally consists of consumer goods such as furniture and computers and, according to FRA, has been the fastest growing segment of the freight rail industry in the United States since 1980.

Inbound international rail traffic has grown over the past 5 years, but the increase is not uniform across U.S. POEs and is projected to increase further in certain POEs. According to BTS data, the number of inbound trains increased 6 percent on the northern border and 29 percent along the southern border from 2010 through 2014.[11] All international rail traffic enters and exits the continental United States through 30 different rail POEs—23 along the Canadian border and 7 along the Mexican border.[12,13] The top 8 rail POEs on the northern and southern borders carried 68 percent of inbound rail traffic while 14 rail POEs—mainly along the northern border—received less than one inbound train a day on average over the past five years according to BTS data (see fig. 1). Ranier, Minnesota, and Laredo, Texas, have the highest number of inbound trains on the northern and southern borders with an average of 10 and 9 trains per day from 2010 through 2014, or an average of 3,675 and 3,466 inbound trains per year, respectively. Some stakeholders predict growth in international rail traffic in certain POEs. For example, representatives from one railroad noted that intermodal traffic through Ranier, Minnesota, will continue to grow since the port at Prince Rupert in Canada has announced an expansion of its capacity. In addition, carmakers announced that they have added additional plants and increased capacity in Mexico, which is likely to result in additional automotive traffic by rail over the southern border.[14]

Train movements can result in blocked highway-rail grade crossings, where vehicular traffic must wait to cross the tracks when trains are slowed or stopped (see fig. 2). The amount of time that highway-rail grade crossings are blocked depends on a number of factors, and is typically a function of the number, speed, and length of trains. Blocked highway-rail grade crossings can contribute to community vehicular congestion, and communities face challenges prioritizing and funding projects to alleviate these impacts. Negative community effects resulting from blocked highway-rail grade crossings include delays to motorists, blocked emergency vehicles, and quality of life impacts.[15] State and local departments of transportation, which have primary responsibility for building, maintaining, and operating roads, can plan and fund projects to alleviate freight-related traffic congestion. In addition, some MPOs assist state and local governments in planning and prioritizing such projects, including grade separation projects such as overpasses and underpasses to allow vehicular traffic to bypass freight rail movements. The freight rail system operates almost exclusively on infrastructure that is owned, built, maintained, and funded by private railroads, particularly the seven largest freight railroads.[16] Generally, train movements within the United States are dispatched, or controlled, by railroad personnel located in the United States.[17]

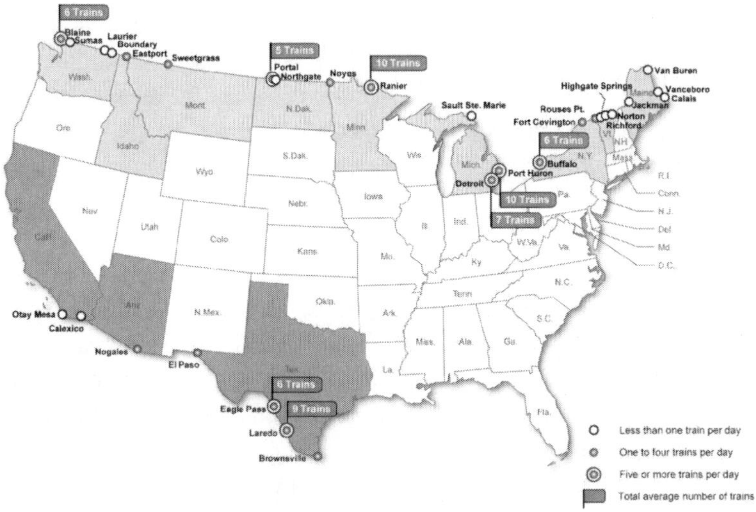

Source: GAO analysis of Bureau of Transportation Statistics data and Map Resources (map). | GAO-16-274.

Figure 1. The 30 Rail Port of Entry Communities and Average Daily Number of Inbound Trains, 2010–2014.

Source: GAO. | GAO-16-274.

Figure 2. A Highway-Rail Grade Crossing in Laredo, Texas.

While DOT has a role in directing federal transportation policy, including freight rail, FRA issues regulations as part of its role to oversee the safety and reliability of the national freight network. In 2012, the Moving Ahead for Progress in the 21st Century Act (MAP-21) transportation reauthorization established a framework for a national freight policy and, among other things, directed DOT to develop a national freight strategic plan.[18] The plan was to be developed in consultation with state departments of transportation and other transportation stakeholders and was to include best practices to mitigate the impacts of freight movements on communities. MAP-21 also required DOT to encourage states to develop freight plans with a description of procedures to guide states' investment decisions involving freight transportation. FRA issues regulations that set requirements for train crews and equipment operating in the United States. Additionally, FRA manages a National Highway-Rail Crossing Inventory that provides a uniform national database of the nation's highway-rail grade crossings, which can be used for planning and implementation of crossing safety improvements. According to the FRA, train lengths in general have been increasing in recent years and agency regulations do not place restrictions on the amount of time trains can block highway-rail grade crossings or on train lengths. Representatives from two railroads noted that current maximum train lengths are generally 10,000 feet—or about 2 miles. These representatives noted that these maximum train lengths are largely determined based on the capacity of the current rail system infrastructure.

As part of its mission to safeguard U.S. borders while enabling legitimate trade and travel, CBP has personnel, including CBP Agricultural Specialists, located at rail POEs that scan inbound trains for security threats. CBP procedures generally include the following, which CBP officials said may vary slightly by POE:

- *Advanced targeting:* About 2 hours before the train arrives at the border, CBP electronically obtains the train's manifest, which provides information on the train's contents, from the railroad. Using CBP's Automated Targeting System, CBP officials identify rail cars deemed high-risk for additional inspection.[19] For example, as part of efforts to identify high-risk shipments, CBP Agricultural Specialists check the manifest against U.S. quarantine regulations.
- *Rail Vehicle and Cargo Inspection System (R-VACIS):* Inbound trains slow to pass through R-VACIS, a machine that produces an image of the inside of railcars using gamma radiation technology (see fig. 3).

CBP officers review the scanned images for anomalies that may indicate the presence of un-manifested goods and contraband, including threats that could pose a risk to national security.

Source: GAO. | GAO-16-274.

Figure 3. R-VACIS in Blaine, Washington, (left) and a train proceeding through R-VACIS in Laredo, Texas (right).

- *Secondary physical inspections:* Depending on the outcome of the advanced targeting and R-VACIS scan, CBP conducts secondary physical inspections of rail cars.

Both DOT and CBP participate in working groups consisting of representatives from the United States, Canada, and Mexico that seek to improve processes related to the safety and fluidity of international trade, including freight rail. Coordination between the United States and Mexico and Canada is generally framed by larger government-to-government partnerships. The U.S.-Canada Beyond the Border Initiative addresses cross border policies and the U.S.-Canada Regulatory Cooperation Council coordinates the joint development of regulatory standards between the United States and Canada, and the High Level Economic Dialogue between Mexican and U.S. officials is designed, in part, to secure trade flows and cross-border cooperation between the two countries. In addition, the Transportation Border Working Group between the United States and Canada and the U.S.-Mexico Joint Working Committee on Transportation Planning focus on transportation issues. For example, the U.S.-Mexico Joint Working Committee on Transportation Planning led efforts to create border master plans to prioritize transportation needs along the southern border, including at rail POEs. To develop these border master plans, local, regional, state, and federal stakeholders on both sides of the border coordinated to prioritize transportation projects.

INSPECTIONS AND CREW CHANGES AFFECT RAIL MOVEMENTS IN SELECTED POE COMMUNITIES, AND SOME ACTIONS HAVE BEEN TAKEN TO EXPEDITE TRAINS

Customs Inspections Affect Train Movements on Both Borders and CBP Has Modified Procedures in Certain Locations

In all four communities we visited, stakeholders such as railroads, local officials, and BLET representatives identified R-VACIS inspection procedures, which affect inbound trains, as a key source of reduced train speeds. CBP has directed that inbound trains pull through the R-VACIS at a predetermined rate of speed set by CBP in order to obtain and review quality scans.[20] The impacts of R-VACIS inspections on train movements and highway-rail grade crossings can vary by the location of the RVACIS. According to CBP officials, the machine is typically located right at the international border, with the exception of three locations on the northern border. The R-VACIS in Blaine is located approximately 3 miles inland from the Canadian border.[21] According to a railroad representative in Blaine, the average maximum length of trains at this POE is 6,500 feet.

Based on our calculations, it would take a train of this length approximately 15 minutes to pass through the R-VACIS at 5 miles per hour and may affect one or two highway-rail grade crossings. In contrast, CBP officials stated that the R-VACIS machines at the Port Huron and Detroit, Michigan, POEs are located in Canada. Trains pass through the R-VACIS in these locations at a predetermined speed and, once scanned, can proceed to enter the United States at a higher speed. CBP officials noted that these placements, which resulted from a Declaration of Principles for the improved security of rail shipments from Canada to the United States, were necessary because the tunnel infrastructure at these POEs requires that trains exit at high speeds. CBP officials also noted that they do not have the authority to physically inspect cargo in Canada.[22]

In addition, when secondary physical inspections occur, they may require trains to slow and stop, and CBP officials stated that the location of the inspections varies by POE and threat level CBP designated to the shipment. CBP officials also said that higher-risk threats, such as shipments containing suspected unauthorized persons (known as stowaways) or weapons, are inspected immediately and that lower-risk threats, such as paperwork discrepancies, are inspected later further away from the border. For example, CBP officials stated that CBP does not use R-VACIS to intentionally scan for people; however, CBP officials in Laredo said that if CBP officers do detect a stowaway on the train, the individual must immediately be secured and removed and could result in the train being stopped for about 45 minutes, during which highway-rail grade crossings on the U.S. side may be blocked. CBP officials in Laredo stated that eight stowaways were inadvertently detected on these trains last year, mostly at night. Meanwhile, more routine secondary physical inspections may involve stopping the train, uncoupling cars, reversing, stopping, and going forward again in order to set aside a rail car for CBP. Depending on the rail infrastructure at the POE, this process may result in trains blocking highway-rail grade crossings.[23] For example, in Blaine, a BLET representative noted that putting a rail car aside for CBP, which generally occurs near the location of the R-VACIS, can take over an hour while blocking highway-rail grade crossings.

As previously mentioned, CBP's primary mission is to maintain national security, and CBP officials report that they operate on risk-based assessments. However, CBP has taken steps to expedite customs inspections at some POEs. CBP officials note that at the POE level, CBP often works together with local communities to develop protocols to expedite rail and minimize the impact on vehicular traffic. In at least two POEs on the northern border, CBP has adjusted the R-VACIS procedures to expedite freight rail. In Blaine, CBP allows empty coal trains through at an increased speed predetermined by CBP during daylight hours unless information received indicates a security risk or there is an operational need, thereby reducing the estimated average blocked highway-rail grade crossing time. In Ranier, a CBP official noted that CBP held meetings to review operations and, as a result, increased the maximum allowable RVACIS speeds to a predetermined rate of speed set by CBP. One CBP official stated that CBP will not sacrifice security for expediency. In addition, at one POE, the railroad coordinated with CBP to expedite secondary inspections. Specifically, in Ranier, railroad officials said that the railroad invested approximately $10 million in equipment, staff, and infrastructure to build a "live lift" system to allow the removal of only the container of interest from intermodal trains for immediate inspection, instead of uncoupling the entire car which could hold several containers (see fig. 4). CBP officials and representatives from the railroad in Ranier stated that this investment reduced the overall secondary physical inspection process time and train delays, as well as the amount of time trains blocked a nearby highway-rail grade crossing.

Source: GAO. | GAO-16-274.

Figure 4. The "Live Lift" System at Ranier, Minnesota.

CBP officials in Laredo and DOT officials stated that trains going into Mexico are also subject to customs inspections, including R-VACIS scans, conducted by Mexican customs officials, which can result in slowed and stopped outbound trains and blocked highway-rail grade crossings in the United States.[24] AAR representatives stated that Mexico is becoming more aware of the need to streamline processes and increase efficiency, particularly now that automobile manufacturing is expanding in Mexico, and U.S. railroads have been working with Mexican officials and other stakeholders to improve processes. For example, AAR representatives said that they meet regularly with customs agencies in the United States, Canada, and Mexico, and that they support a Trans-border Committee comprised of member railroads from all three countries to promote simplification and the development of electronic reporting systems to expedite freight rail traffic. At the POE level, CBP officials do not have authority over train movements once trains have crossed the border into Mexico or Canada.[25]

Brake Inspections Affect Inbound Trains on the Southern Border, and FRA Has Waived Certain Requirements to Expedite Trains

Trains entering the United States from Mexico must stop at the border for FRA-required brake inspections, and FRA has waived certain requirements to expedite this process.[26] FRA regulation requires crews to perform full brake tests on trains at the origin location or at the interchange point, which is generally at the border as the trains enter the United States.[27] An FRA region official stated that full brake tests were previously conducted with the whole train on the U.S. side, which could block highway-rail grade crossings for up to an hour. These brake tests include performing an air leakage test to ensure air brake pressure is maintained throughout the train, as well as a visual inspection of each car's air brakes.[28] Since the early 2000s, FRA has granted waivers to railroads to conduct abbreviated brake inspections at the border, provided the railroad submits a waiver request that meets certain criteria and is consistent with railroad safety. U.S. railroads on the southern border now have FRA brake inspection waivers in all but one POE, and FRA officials and railroad and BLET representatives said that such waivers to allow abbreviated brake tests have resulted in expedited train movements.[29]

The abbreviated brake tests allowed through the waiver can take 20 to 25 minutes according to BLET representatives in Laredo. An abbreviated brake test requires a visual roll-by inspection and a set-and-release test of the air brakes where the crew uses an end of train device to ensure air pressure is reaching the end of the train.[30] As a condition of the waiver, crews are then required to conduct a full brake inspection at a U.S. rail yard away from the border.

Despite FRA's efforts to expedite brake inspections along the southern border, inbound trains sometimes arrive from Mexico with missing or damaged equipment which can cause delays. According to BLET and railroad representatives in Laredo, trains from Mexico often arrive in the United States with missing "end-of-train devices" that are required for the abbreviated brake test, which can cause delays up to an hour as train crews locate a replacement device. In addition, railroad and BLET representatives in Laredo noted that it is common for other train equipment to be tampered with, a situation that requires the train to be stopped until repairs can be completed.

The Rail Safety Improvement Act of 2008 prohibits FRA from accepting mechanical and brake inspections of rail cars performed in Mexico before entering the United States unless, among other criteria, FRA certifies that the inspections are being performed under regulations and standards equivalent to those applicable in the United States.[31] Moreover, according to DOT officials, FRA officials cannot verify brake inspections conducted in Mexico, in part, because the FRA officials face challenges coordinating with their counterparts due to security concerns.[32] As a result, brake inspections occur on the border between the United States and Mexico, typically on a bridge. According to DOT officials, greater harmonization between the pertinent U.S. and Mexican regulations could result in the United States' accepting brake inspections conducted in Mexico. DOT officials noted that although they would like to discuss rail regulatory and safety issues with Mexico and considers rail-related issues on occasion, no rail regulation harmonization efforts are currently underway, in part because Mexico is currently restructuring its rail regulatory body in an effort to increase its rail investments and networks. Furthermore, the U.S.-Mexico working group's coordination efforts such as the U.S.-Mexico Joint Working Committee on Transportation Planning, have had limited initiatives focused specifically on freight rail issues, having instead focused on issues facing passenger vehicles and freight trucks. As we have previously mentioned, 60 percent of the freight that moves between the United States and Canada and Mexico is carried by truck.

Crew Changes Affect Inbound and Outbound Trains on the Southern Border due to Factors Such as Differing Safety Regulations

DOT officials told us that inbound and outbound trains on the southern border are required to stop at the border to change crew due to lack of comparable rail safety regulations between the United States and Mexico.[33] While a BLET representative stated that crew changes can take 3 to 5 minutes, this can vary greatly depending on crew availability. For example, BLET and railroad representatives in Laredo noted that crews, who deliver trains to the rail yard and then are driven by a rail crew van to the border to pick up another train, can get delayed at the yard or on the way back to the border by traffic congestion. Such delays, according to a BLET representative in Laredo, can result in crew changes exceeding 2 or 3 hours. FRA regulations establish minimum federal safety standards for the eligibility, training, testing, certification, and monitoring of all locomotive engineers and conductors.[34] According to DOT officials, the lack of Mexican safety regulations for the qualification and certification of locomotive engineers and conductors that are comparable to FRA regulations prohibits the United States from allowing Mexican crews to operate trains in the United States. In addition, as previously mentioned, while greater regulatory harmonization could result in Mexican crews being able to operate in the United States, DOT officials noted that Mexico is currently focused on creating a rail transport regulatory agency. According to DOT, FRA will invite Mexico to attend the annual North American Rail Safety Working Group Meeting in 2016 in an effort to encourage further harmonization.

Two railroads have expressed interest in developing an international pool of crew to eliminate the need for crew changes on the southern border; however, DOT and CBP officials, and BLET representatives cited barriers to this initiative. Specifically, DOT officials stated that qualification and certification regulations, varying operating rules and hours of service for crews, and labor and union concerns would need to be addressed. Additionally, CBP officials in Laredo stated that they do not currently have the capability needed to facilitate processing an international crew.[35] BLET representatives also noted concerns such as liability for damages and personal injury and security if U.S. crews were to operate in Mexico, since federal workplace laws are not applicable to U.S. citizens injured on the job while working abroad.[36] BLET representatives also noted concerns with personal security of crew members while on board the train or when returning to the United States by vehicle after delivering the train to its destination in Mexico. These representatives also noted that exceeding the federal maximum allowable hours of service might become an issue given delays re-entering the United States at the vehicle border crossing.[37]

CBP and FRA have limited information on the effects of the above factors on rail movements. Although CBP has personnel located at the border, it does not have visibility into all factors affecting train movements. For example, trains are often operated at restricted speeds through POEs, meaning speeds are dictated by factors such as the train's stopping distance and the train operator's range of vision. According to BLET representatives in Ranier, speeds can be anywhere from 0.5 to 10 miles per hour through town due to the long stopping distances of heavy trains combined with limited visibility as a result of factors such as inclement weather or the track curvature, regardless of factors such as CBP inspections. Meanwhile FRA, which is primarily focused on the safety of trains operating within the United States, does not have staff located at POEs. Instead, FRA officials stated that they rely on voluntary reporting from railroads on any delays occurring and the reasons for these delays. FRA officials noted that it is difficult to obtain data from railroads on the cause and extent of train-related delays in POEs. CBP and FRA officials stated that they rely on communication with stakeholders to inform decisions such as modifying CBP procedures or brake test waiver requirements. As discussed later in this report, FRA has undertaken efforts to improve the availability of data on freight rail movements, including those at POEs.

INTERNATIONAL FREIGHT RAIL IMPACTS VARY BY COMMUNITY GAO VISITED, AND DOT'S DATA IMPROVEMENT EFFORTS COULD HELP DETERMINE THE EXTENT OF BLOCKED HIGHWAY-RAIL GRADE CROSSINGS

Impacts of International Freight Rail on Communities GAO Visited Vary Based on Border-Specific Factors and Community Characteristics

The factors noted above—customs inspections, brake inspections, and crew changes—can slow or stop trains travelling through U.S. POEs and consequently block highway-rail grade crossings in those communities, but different POEs are affected differently. As noted in Figure 5, the effect of factors such as customs inspections can vary based on whether the community is located on the southern or northern border. For example, an outbound crew change can result in the train stopped in one or more highway-rail grade crossings on the southern border, but is less likely to occur on the northern border because of greater harmonization, among other factors, between U.S. and Canadian safety regulations. In addition, although U.S. customs inspections can block U.S. highway-rail grade crossings for inbound trains on both borders, foreign customs inspections primarily impact outbound trains on the southern border.

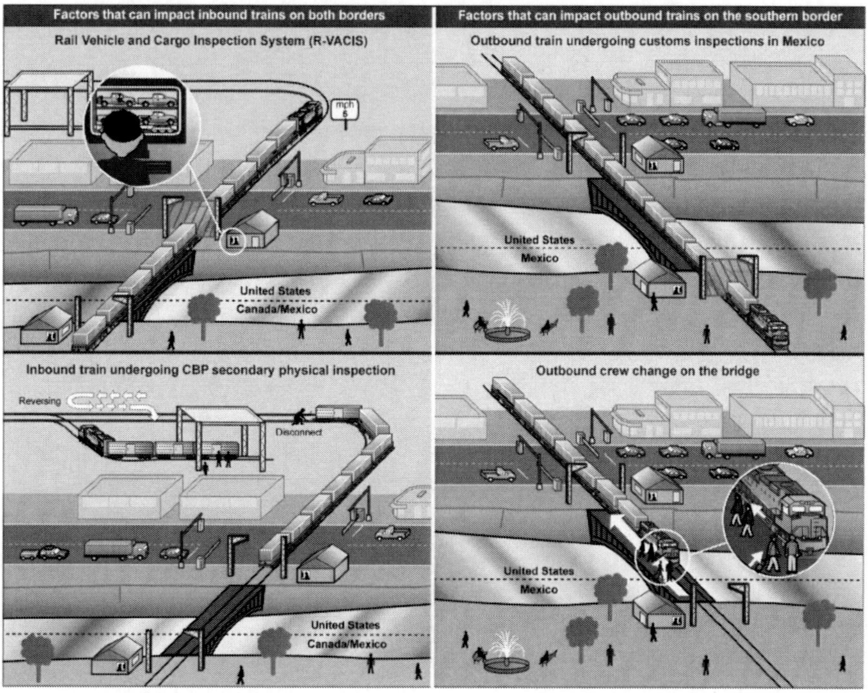

Source: GAO. | GAO-16-274.

Figure 5. Examples of Factors That Can Affect the Time That Highway-Rail Grade Crossings Are Blocked in U.S. Port of Entry Communities.

The extent to which the above factors may result in a train blocking a highway-rail grade crossing and delaying vehicular traffic also vary due to community characteristics, such as the number and location of highway-rail grade crossings and the availability of overpasses. For example, as noted below, in Ranier, railroad representatives estimated that one key highway-rail grade crossing is blocked for about 8 hours per day. In contrast, MPO officials in Buffalo and Detroit reported that international freight rail movements have minimal impact on traffic congestion in those cities because the rail lines are largely grade-separated, meaning the rail line rarely intersects with vehicular traffic.

Furthermore, we have previously found that although communities may have long-standing concerns with the negative effects of highway-rail grade crossings, they have varying levels of quantified information on impacts such as traffic delay times or costs.[38] Similarly, POE communities we visited provided some estimates of the amount of time highway-rail grade crossings are blocked, but were unable to provide data on the actual extent of blockage. For example, local officials in Blaine note that hour-long traffic disruptions can result from blocked highway-rail grade crossings, with 30 minutes waiting for the train and another 30 minutes waiting for the vehicle traffic queue to clear. However, local officials reported they did not have information on how regularly such delays occurred due to a lack of data.

The following discussion of the rail POE communities we visited illustrates how their characteristics impacted highway-rail grade crossings.

- **Ranier, Minnesota:** Ranier is a community of 145 according to the 2010 Census, and is located about 3 miles northeast from the larger community of International Falls, Minnesota. Within Rainer, there is one highway-rail grade crossing—Spruce Street (see fig. 6).

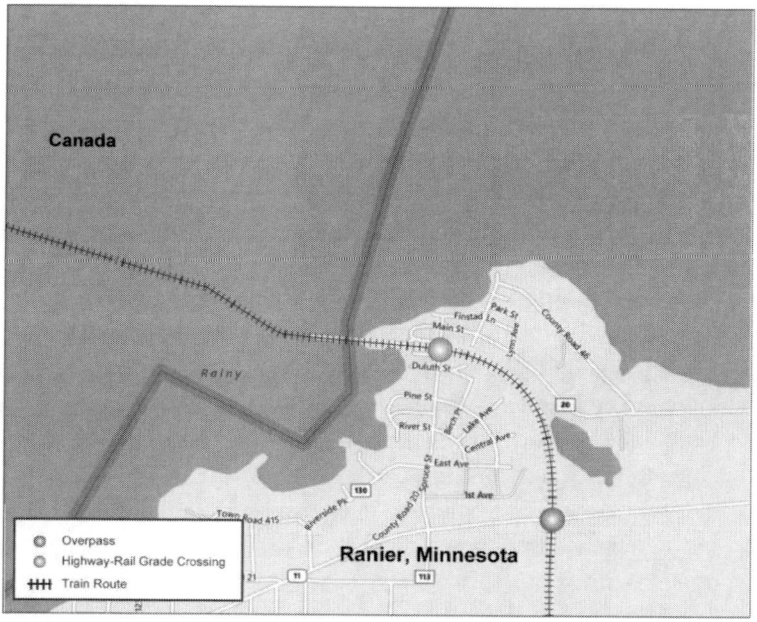

Sources: GAO analysis of Federal Railroad Administration data and MapInfo. | GAO-16-274.

Figure 6. At-Grade and Grade Separated Highway-Rail Crossings in Ranier, Minnesota.

Spruce Street is blocked about 8 hours per day by the 20–22 trains traveling through per day—about 11 in each direction—according to representatives from the railroad. These representatives arrived at this total by estimating that a southbound train takes about 25 minutes to pass the highway-rail grade crossing, and a northbound train takes about 15 minutes, which amounts to over 7 hours a day for 11 trains to pass in each direction. These representatives report that the train traffic is distributed across nighttime and daytime hours because of the railroad's aim to move traffic over its network evenly, which results in about one train travelling through Spruce Street per hour, including through the night. Speeds are slowed for inbound trains through Spruce Street due to CBP's R-VACIS, although, as mentioned previously, CBP has taken efforts to expedite R-VACIS and the railroad and CBP have worked together to implement the live lift system to expedite secondary inspections. According to local officials, the blockage of Spruce Street has had a debilitating effect on businesses located north of Spruce Street. These officials report that due to the proximity of the Spruce Street intersection to Rainy Lake, it is impossible to build an overpass at that location. However, an overpass located approximately a mile away helps vehicle traffic reroute to get around the train. According to an FRA region official, the situation in Ranier does not constitute a serious effect on vehicle traffic, particularly compared with POE communities on the southern border and given the presence of the overpass.

- **Blaine, Washington:** Blaine, which is 35 miles south of Vancouver, Canada, is bordered on the north by the U.S./Canada border. The community—population 4,684 according to the 2010 Census—includes both Central Blaine to the east and West Blaine, where the Semiahmoo resort and marina are located. The rail line is located close to the waterfront through Central Blaine. Local officials report that two key highway-rail grade crossings are affected by freight rail movements— Hughes Avenue, a sole access point to a neighborhood of approximately 300 residents; and Bell Road, a key route connecting Central Blaine to West Blaine's resort and marina (see fig. 7).

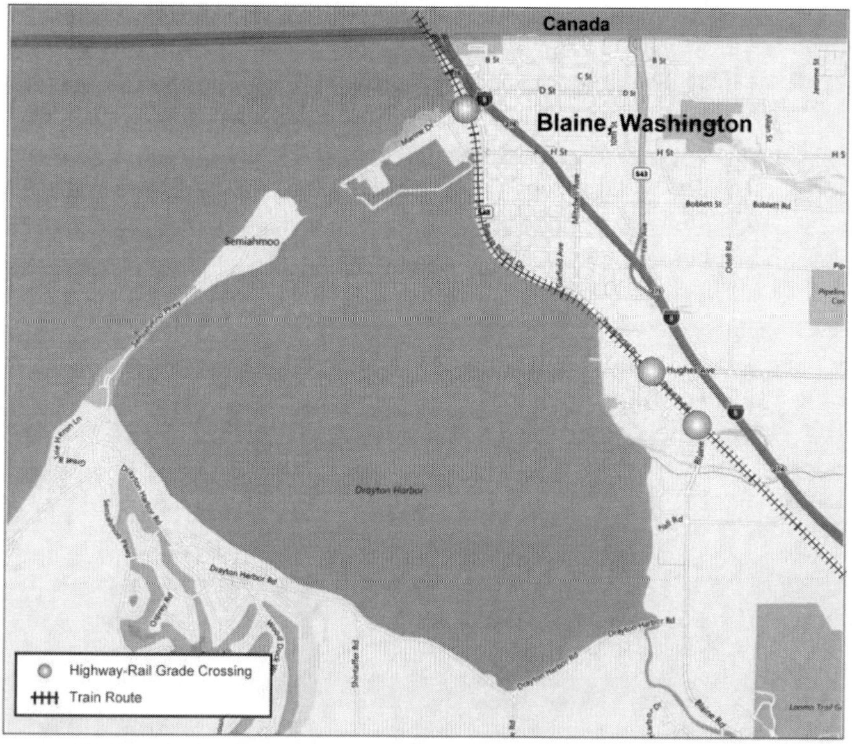

Source: GAO analysis of Federal Railroad Administration data and MapInfo. | GAO-16-274.

Figure 7. Highway-Rail Grade Crossings in Blaine, Washington.

According to railroad representatives, 12 freight trains pass per day— 6 in each direction—through Blaine, at both day and nighttime hours.[39] Local officials attribute issues related to blocked highway-rail grade crossings in Blaine to the R-VACIS; however, as mentioned previously, CBP has adjusted its procedures to enable certain trains to go through R-VACIS faster. Local officials were unable to provide data on the amount of time Hughes Avenue and Bell Road are blocked, and noted that it is difficult to fund traffic studies that take train traffic into account, in part because the railroad does not contribute funding. Within Blaine there are no overpasses to enable traffic to reroute around trains. Furthermore, local officials reported it is not feasible to construct overpasses over Hughes Avenue and Bell Road due to geographic limitations such as the location of homes and a creek.

- **Laredo, Texas:** The 2010 census reported that Laredo is a city of approximately 236,000, and every day about 22 trains travel through Laredo—11 inbound and 11 outbound, according to CBP officials. Information provided by one of the railroads indicates that this traffic is fairly evenly split between daytime and nighttime hours. According to a 2006 study prepared for the MPO and the city, Laredo has over 80 highway-rail grade crossings which are split fairly evenly between two rail lines, which are operated by two different railroads and carry traffic in different directions through the city. A railroad representative noted that train traffic has recently been evenly split between these two rail lines. One of these rail lines bisects the downtown area, with 13 at-grade highway-rail crossings located at about every block (see fig. 8).

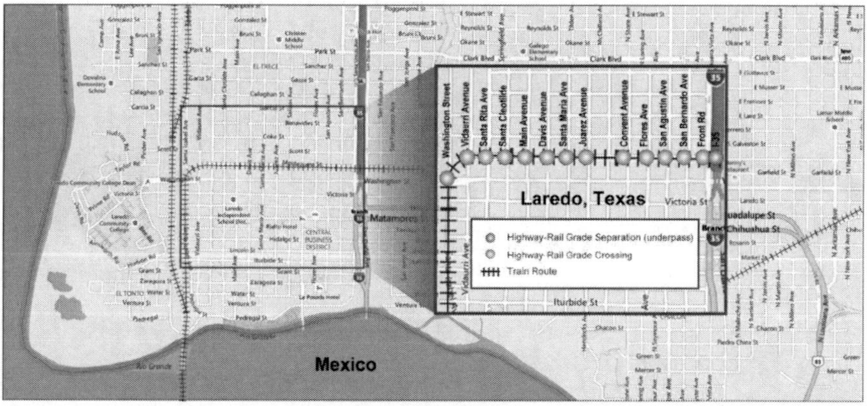

Source: GAO analysis of Federal Railroad Administration data and MapInfo. | GAO-16-274.

Figure 8. Selected Highway-Rail Grade Crossings in Laredo, Texas.

According to an MPO official, the majority of complaints regarding blocked highway-rail grade crossings are along this downtown portion of the rail line. CBP officials in Laredo noted that a single stopped train can stretch from the border to near Interstate 35, a distance of approximately 2 miles, blocking all of the highway-rail grade crossings in between, including the 13 located downtown. These officials noted that this can affect traffic downtown, including lawyers who are cut off from the federal courthouse located on the other side of the rail line from their offices. In 2012, the Laredo region developed a Border Master Plan, which convened local, regional, and federal officials on both the U.S. and Mexico side of the border to prioritize border transportation projects. According to Texas state DOT officials, the Border Master Plan demonstrated the need for accurate data, including on current and future vehicular traffic levels, for analyzing costs and benefits and prioritizing projects. In addition, in 2015, a Laredo MPO-commissioned study gathered data on the number of trains passing through the community and speed from the Highway Rail Crossing Inventory, as well as vehicular traffic counts. However, since this study was primarily focused on actions to reduce train horn noise, it did not calculate the total amount of time highway-rail grade crossings are blocked.[40]

- **Brownsville, Texas:** A community of about 175,000 people according to the 2010 Census, Brownsville currently has about 4 to 8 trains pass through the community per day, according to a railroad representative. On August 25, 2015, the first new international rail crossing between the United States and Mexico in 105 years was inaugurated in Brownsville. The new rail bridge relocates rail traffic away from the downtown area to the outskirts of Brownsville, with only one highway-rail grade crossing, and eliminates 14 highway-rail grade crossings downtown. Although moving the rail line outside of town has been discussed in other southern rail POE communities such as El Paso and Laredo, only Brownsville has succeeded in moving the rail POE out of the downtown area.[41] A Cameron County official noted that project planning began in the 1990s, that much of the data used to prioritize the project was taken from a detailed feasibility study, and that other communities should now have an easier time proposing similar projects given that states are more involved with freight rail planning. According to a county official, the U.S. portion of the project cost over $40 million and most of the funding came from federal sources, including the American Recovery and Reinvestment Act of 2009.[42] According to a railroad representative, the railroad agreed to transfer a portion of its existing right of way land to the county in exchange for the new right of way and infrastructure constructed by the county. Therefore, the railroad's contribution to the project was the value of the land exchange rather than directly contributing funding for the new construction. In addition, a county official noted that coordinating with officials from Mexico and CBP were key challenges. Specifically, this official noted that monitoring the progress of the project on the Mexican side and

> coordinating with CBP on its requirements for the new bridge, such as the relocation of R-VACIS, posed challenges. CBP officials in Brownsville noted that the project did not begin with good coordination, and cited the need for strong coordination as a "lesson learned." CBP, FRA region, and Brownsville MPO officials noted that the long-term success of the new rail bridge will largely depend on development of the area.[43] These officials stated that increased development may result in new highway-rail grade crossings, which could result in traffic issues over time. A railroad representative noted that rail traffic through Brownsville is expected to increase in the future.

The effect that freight rail may have on communities also varies based on the time of day that trains pass through the rail POE communities, as well as efforts made by railroads to prevent trains from blocking certain highway-rail grade crossings. For example, as noted above, trains pass through Ranier, Minnesota, around the clock, at an average of one per hour according to railroad representatives. Therefore, about half of the trains run through at night, when vehicle traffic is less and traffic congestion not an issue. In addition, according to railroad representatives and MPO officials in El Paso, trains cross the border during night time and early morning hours due to a Juarez, Mexico, city ordinance that restricts train movements to those times. In some situations, railroads have worked to avoid blocking certain highway-rail grade crossings. For example, in Laredo, a railroad representative noted that crews make best efforts to avoid blocking a trucking route and street with a school nearby during school hours. In addition, in Blaine, a CBP official reported that the railroad tries to limit the number of trains going through the community during the morning rush hour to avoid delaying school buses.

DOT's Data Improvement Efforts May Help Determine Extent of Blocked Highway-Rail Grade Crossings in Rail POE Communities

We have previously found that a lack of publicly available data on freight rail movements and estimates of their impacts on vehicular traffic in communities across the United States creates difficulties in defining the extent of the problem and prioritizing potential solutions.[44] Specifically, we found that limitations in both national and state and local data on freight rail movements reduce the ability of state or local officials to quantify freight rail community impacts nationwide and that these limitations create challenges to appropriately prioritizing efforts to address freight rail impacts against other types of funding priorities. At the national level, data on freight-related traffic congestion for local communities have limitations in terms of timeliness and completeness. At the local level, communities have limited data such as the number of trains and length of trains assigned by date, speed, and time. As we have previously found, communities often find it difficult to communicate with the railroad industry to obtain information on the number, timing, and speed of trains.

We requested data directly from the railroads in order to quantify the extent that freight rail movements blocked highway-rail grade crossings in a selection of rail POE communities. Specifically, we requested data on the number of trains, the length of trains, and the speed of trains from railroads that operate in these POEs. This information would allow us to estimate train blockage time at highway-rail grade crossings in these communities. However, although we requested data directly from the five railroads that operate in eight selected rail POE communities,[45] we received complete information from two of the railroads.[46] Based on this data, we calculated the time selected highway-rail grade crossings are blocked and found highway-rail grade crossings in two communities— Ranier and one of the two rail lines in Laredo—to be blocked on average 16-19 minutes per train.[47]

Recent DOT efforts could help improve the availability of freight rail data needed to assess community impacts such as blocked highway-rail grade crossings for communities across the country, including POE communities. FRA maintains the National Highway-Rail Crossing Inventory that includes information such as the estimated number of daily trains in communities and the typical range of speed of trains that pass through a highway- rail grade crossing. However until recently this information was voluntarily submitted by railroads and states and according to FRA officials was not always current. On January 6, 2015, FRA issued a final rule requiring railroads to update the inventory once every 3 years.[48] FRA officials said that the rule should improve the quality of the data, but that these improvements will not be fully evident for several years. Improved information on the average number of daily trains could better equip state and local governments to identify community congestion impacts from freight rail—including blocked highway-rail grade crossings located in POE communities along the border. Furthermore, in a November 2015 letter to congressional committees regarding a surface transportation bill, DOT Secretary Anthony Foxx noted that given the concerns regarding blocked crossings in many communities, FRA would benefit from authorization and funding to study blocked crossings to collect information as to the severity, frequency, and other characteristics of railroad operations that block highway-rail grade crossings. Secretary Foxx also noted that neither the House or Senate versions of the bill propose such authorization and funding. On December 4, 2015, President Obama signed into law the Fixing America's Surface Transportation Act, which did not contain such provisions regarding blocked crossings.[49]

In addition, in September 2014, we issued a report on freight-related community impacts and recommended, among other things, that DOT incorporate additional information to help states define and prioritize local community impacts of national freight movements, including traffic-congestion impacts, and to establish what data could be consistently collected and analyzed in order to prioritize impacts of freight on local traffic congestion in its final guidance to states in the development of their state freight plans.[50] We also recommended that DOT include a strategy for improving the availability of national data needed to quantify, assess, and establish measures of freight trends and impacts on local traffic congestion for inclusion in its National Freight Strategic Plan. DOT agreed with our recommendations. On October 18, 2015, DOT issued a draft *National Freight Strategic Plan* for public comment. The draft noted that DOT should work closely with state and local governments and international partners, as well as private stakeholders, to coordinate strategies and investments and noted that new freight traffic data sources and improved public-private cooperation on state freight plans will assist in this effort. The draft also noted that DOT should continue to engage in strong border infrastructure planning with border states through working groups with Canada and Mexico. We will continue to monitor the status of DOT's response to our recommendations and DOT's efforts related to the *National Freight Strategic Plan.* A DOT strategy on data to prioritize the impacts of freight related traffic congestion in the *National Freight Strategic Plan*, along with improvements to the National Highway-Rail Crossing Inventory, could help address data limitations at both the national and local levels and help communities—including POE communities—better define impacts from blocked highway-rail grade crossings and prioritize projects to mitigate such impacts.

AGENCY COMMENTS

We provided a draft of this report to DOT and CBP for review and comment. DOT highlighted efforts to minimize community impacts of international freight rail movement. DOT and CBP provided technical comments, which we incorporated.

Susan A. Fleming
Director, Physical Infrastructure Issues

APPENDIX I: OBJECTIVES, SCOPE, AND METHODOLOGY

This report (1) describes factors that affect the movement of freight rail through selected ports of entry and the actions taken by federal agencies and others to expedite freight rail in these locations, and (2) examines what is known about the impacts of freight rail operations on highway-rail grade crossings in U.S. port of entry communities.

To determine the factors that affect the movement of freight rail through selected ports of entry and the actions taken to expedite freight rail in these locations, we interviewed officials and reviewed documents from Customs and Border Protection (CBP), the U.S. Department of Transportation (DOT), the Federal Railroad Administration (FRA), and Department of State. We also interviewed representatives from the American Association of State Highway and Transportation Officials, the Border Trade Alliance, the Association of American Railroads, and the Brotherhood of Locomotive Engineers and Trainmen (BLET)—a union which represents train operators that we identified from prior GAO work. We interviewed FRA officials and reviewed FRA documentation regarding crew changes and brake inspections, including applicable regulations and FRA waiver decisions regarding brake inspections. We also interviewed DOT, FRA, and CBP officials and reviewed documentation on international working groups involving transportation issues on both the northern border (i.e., the U.S.- Canada Regulatory Cooperation Council and the Transportation Border Working Group) and the southern border (i.e., the U.S.-Mexico High Level Economic Dialogue and the U.S.-Mexico Joint Working Committee on Transportation Planning). To determine what is known about the impacts of freight rail operations on highway-rail grade crossings in U.S. POE communities, we also reviewed previous GAO reports and recommendations and interviewed DOT officials on available data sources and reviewed relevant documentation, such as the reporting requirements for the National Highway-Rail Crossing Inventory.

To determine the factors that affect the movement of freight rail and the impacts of freight rail operations on highway-rail grade crossings, we selected nine rail POE communities— Nogales, Arizona; El Paso, Eagle Pass, Brownsville, and Laredo, Texas; Blaine, Washington; Ranier, Minnesota; Port Huron, Michigan; and Rouses Point, New York. These communities were selected because they had at least one inbound train on average per day from 2010 through 2014, according to DOT's Bureau of Transportation Statistics' (BTS) Border Crossing data. As part of this selection, we excluded 11 communities where the rail POEs were in transit (where trains pass through but are not subject to full CBP procedures), outside of the continental United States, did not cross incorporated communities, or have largely grade-separated infrastructure.

We conducted visits to four of these selected communities—Brownsville and Laredo, Texas; Ranier, Minnesota; and Blaine, Washington—that were selected based on factors such as those with heavy inbound train volume from 2010 through 2014 according to BTS data, complaints received by CBP about blocked crossings, and a mix of northern and southern border locations. We also selected locations where actions had been taken to mitigate congestion or expedite rail, such as Brownsville, Texas, for its construction of a new international rail bridge. At each of the four site visits, we interviewed representatives from the city or county, the Metropolitan Planning Organization (if applicable), the state department of transportation, the FRA regional office, and BLET. We also interviewed representatives from the 5 railroads that operate trains through each selected POE. In each site visit we also interviewed officials from CBP and observed their inspection process as well as the geography and relevant highway-rail crossings of the community. We calculated the average time that freight trains would block key highway-rail grade crossings in selected communities based on the average speed of trains, length of trains, and frequency of trains that were reported by railroad representatives. To do so, we developed a data collection instrument and attempted to collect information from five railroads[1] on the number, length, and speed of trains passing over the three highway-rail grade crossings closest to the international border on a typical weekday in July 2015 in eight of the selected communities. [2] As we note in the report, although we requested information from five railroads, we received incomplete information in response and were able to analyze information from two of these railroads.[3] In order to better understand the impacts of international rail in these communities, we spoke to local officials from the city or MPO by phone in each of the five selected communities that we did not visit (Nogales, Arizona; El Paso and Eagle Pass, Texas; Port Huron, Michigan; and Rouses Point, New York). We also interviewed officials from the MPOs in Detroit, Michigan and Buffalo, New York, to understand the impacts of international freight rail in these communities.

We developed maps to provide context regarding the level of international freight rail traffic and impacts on communities. Specifically, we used BTS data to calculate the average number of inbound trains per day from 2010 through 2014 by POE and displayed this information on a map. To determine the reliability of BTS data, we reviewed related documentation and interviewed knowledgeable agency officials. We determined these data were sufficiently reliable for our purpose of providing contextual information. We also developed maps including the location of at-grade and grade separated highway-rail crossings for three of the four communities we visited—Ranier, Minnesota; Laredo, Texas; and Blaine, Washington. We did not include a map of Brownsville, Texas, since its rail traffic patterns are currently changing due to the construction of a new international rail bridge. To develop these maps, we used data from the National Highway-Rail Crossing Inventory, as well as maps and observations obtained from our in-person visits to these communities. By reviewing related documentation, interviewing knowledgeable DOT officials, and comparing the data to our site visits, we determined the data were sufficiently reliable for the purpose of developing maps.

End Notes

[1] The BTS does not collect data on outbound trains. However, trains also leave the United States through these same POEs. This 30 excludes Warroad and Baudette, Minnesota, which are in transit POEs, meaning that trains pass through but do not stop for inspection in the U.S. This also excludes Skagway, Alaska, because it is outside the continental U.S.

[2] Passenger trains pass into the U.S. through three northern POEs: Blaine, Washington; Buffalo-Niagara Falls, New York; and Champlain-Rouses Pt., New York. Amtrak runs 2 inbound trains a day through Blaine and 1 inbound train per day at the two New York POEs.

[3] GAO, *Freight Transportation: Developing National Strategy Would Benefit from Added Focus on Community Congestion Impacts.* GAO-14-740 (Washington, D.C., Sep. 19, 2014).

[4] H. R. Rep. No. 113-464 accompanying Pub. L. No. 113-235, 128 Stat. 2130 (2015).

[5] BTS does not collect data on outbound trains.

[6] Metropolitan planning organizations (MPO) are federally mandated entities responsible for carrying out the metropolitan transportation planning process in urbanized areas with a population of more than 50,000 people. (23 USC 134).

[7] We also interviewed officials from MPOs in Detroit, Michigan, and Buffalo, New York, to understand the impacts of international freight rail in these communities.

[8] Brownsville was excluded because at the time of our visit in late June to early July 2015, the new international rail bridge was nearing completion, and as a result, the railroad was in the process of changing its travel pattern, making it difficult to characterize the impacts of freight rail on the community.

[9] These railroads were: Kansas City Southern Railway Company, Union Pacific Railroad Company, BNSF Railway Company, Canadian National Railway Company, and Canadian Pacific Railway.

[10] We received information from three railroads, but information from one of these railroads was incomplete.

[11] BTS does not collect data on outbound trains. However, railroad representatives in the four POEs we visited noted that the same number of trains travel inbound as outbound in those locations on a typical day.

[12] This 30 excludes Warroad and Baudette, Minnesota, which are in transit POEs, meaning that trains pass through but do not stop in the U.S., and thus are not subject to full CBP inspections. This number also excludes Skagway, Alaska, because it is outside the continental U.S. In some cases, the official POE name differs from the name of the U.S. community with the international rail line. For the remainder of this report we will refer to the name of the rail POE communities rather than the POE name. As a result, we refer to the International Falls POE as Ranier, Minnesota; the Pembina, North Dakota POE as Noyes, Minnesota; the Buffalo-Niagara Falls POE as Buffalo, New York; the Champlain-Rouses Pt. POE as Rouses Pt., New York; and the Trout River/Fort Covington/Chateaugay POE as Fort Covington, New York.

[13] According to BTS data, there were 88 POEs where at least one truck per day entered the continental United States in 2014.

[14] Audi of America, Inc., *Audi on track for growth in Mexico,* (Ingolstadt, Germany: Jan. 22, 2014); The BMW Group, *BMW group to build plant in Mexico,* (Munich, Germany and Mexico City, Mexico: Mar. 7, 2014); General Motors Co., *GM to Invest $5 billion in Mexico from 2013-2018*, (Federal District, Mexico: GM News, Dec. 11, 2014); Honda, *Honda Increases North American Manufacturing Footprint with Production Start of Fuel-Efficient, Subcompact Vehicles at New Auto Plant in Mexico,* (Celaya, Mexico: Feb. 21, 2014).

[15] GAO-14-740.

[16] These railroads are referred to as Class I railroads. Freight railroads are classified based on operating revenues. Class I railroads have annual operating revenues of $467 million or more. As of 2013, AAR reported that the seven Class I railroads are BNSF Railway Company, CSX Transportation, Grand Trunk Corporation, Kansas City Southern Railway Company, Norfolk Southern Combined Railroad Subsidiaries, Soo Line Corporation, and Union Pacific Railroad Company.

[17] See 49 C.F.R. § 241.9—Prohibition against extraterritorial dispatching; exceptions.

[18] Pub. L. No. 112-141, §1115, 126 Stat. 405, 468. 23 U.S.C. § 167(f).

[19] CBP's Automated Targeting System is an Intranet-based enforcement and decision support system that compares traveler, cargo, and conveyance information against intelligence and other enforcement data.

[20] According to a 2004 CBP report, R-VACIS can scan moving freight train rail cars with a speed up to 5 miles per hour.

[21] CBP officials say the inland location of R-VACIS in Blaine is due to building restrictions on protected land near the border.

[22] According to CBP officials, if CBP officers want to physically inspect a train, they notify CBP officers in the United States to conduct the inspection upon its arrival. These officials also stated that in order to physically inspect cargo in Canada, CBP would require greater authority than that provided by the signing of a Declaration of Principles with Canadian Customs which requires legislative approval in both countries to go into effect.

[23] According to a CBP Laredo official, as of October 2015, routine physical inspections at this POE are being conducted at the railroad's secondary exam station or warehouse. According to this official, only immediate threats result in stopped trains at the rail POE crossing.

[24] CBP officials and railroad representatives at the two northern border POEs we visited stated that Canada does not use R-VACIS to scan inbound trains, and does not stop trains at the border for inspections. As a result, trains generally leave the United States at unimpeded speeds on the northern border. For example, a railroad representative in Blaine reported that outbound trains go through Blaine at a minimum of 45 miles per hour.

[25] We did not speak with Mexican or Canadian customs officials for this report.

[26] Railroads must submit a waiver petition to FRA for consideration, and FRA will publish a notice seeking public comment and may conduct a field investigation or a public hearing if necessary. If FRA determines to grant a waiver, such waivers last for up to 5 years and may be renewed upon request.

[27] On the northern border, according to DOT officials, FRA accepts brake inspections conducted in Canada due to greater harmonization of FRA regulations with Canadian regulations and strong similarities in safety requirements.

[28] 49 C.F.R. § 232.205 *Class I Brake test-initial terminal inspection* states that each train and each car in the train will receive a Class I brake test by a qualified person, who has the required training, qualification, designation, and instruction to perform such functions. Throughout this report we refer to Class I brake tests as full brake tests.

[29] FRA has issued brake waivers for both of the southern POEs we visited—Laredo and Brownsville, Texas.

[30] 49 C.F.R. § 232.211 *Class III Brake tests-trainline continuity inspection.* Throughout this report we refer to Class III brake tests as abbreviated brake tests. An "end-of-train device" is a portable electronic device placed at the end of freight trains to monitor air brake pressure.

[31] Under Pub. L. No. 110-432 § 416, 122 Stat. 4890 (2008) as codified in 49 U.S.C. § 20107. For brake tests to be accepted from Mexico, inspections must meet certain criteria that are certified by the Secretary of Transportation.

[32] The Department of State places travel restrictions on U.S. government employees in Mexico. U.S. government employees are subject to movement restrictions and a curfew between the hours of midnight and 6 a.m. in the Mexican state of Tamaulipas due to violent crime. This includes Matamoros and Nuevo Laredo, which are the cities adjacent to Brownsville and Laredo, respectively.

[33] FRA stated that crew changes are not mandatory on the northern border as the safety and qualification regulations and labor unions in Canada more closely resemble those in the United States. Of the two locations on the northern border we visited, only crews in Ranier changed at the border, which railroad representatives noted was in part for logistical and transportation considerations. Ranier city officials noted that eliminating crew changes

could increase speeds and reduce the amount of time Ranier's one highway-rail grade crossing is blocked. However, railroad representatives noted that eliminating crew changes, which do not result in stopped trains blocking this highway-rail grade crossing, would have a minimal impact on speeds at this location.

[34] 49 C.F.R. Parts 240 and 242 *Qualification and certification of locomotive engineers and conductors*.

[35] CBP officials stated that crews from Mexico and Canada require proper admissibility documents to enter the United States.

[36] Federal Employers Liability Act c. 149, 35 Stat.65 (1908) codified as amended in 45 U.S.C. § 51, *New York Central Railroad Company v. Chisholm, Administrator, 268 U.S. 29 (1925).*

[37] 49 U.S.C. §21103 set the hours of work and rest of train employees.

[38] GAO-14-740.

[39] In addition, according to the state DOT, 4 passenger trains pass through Blaine per day—2 northbound and 2 southbound. This Amtrak route runs from Oregon to Vancouver, Canada. However, according to local officials, passenger trains travel through Blaine at higher speeds than freight trains and are less of an issue in terms of blocked highway-rail grade crossings.

[40] Under the train horn rule, locomotive engineers must begin to sound train horns at least 15 seconds in advance of all public highway-rail grade crossings. The rule also provides an opportunity for communities to mitigate the effects of train noise by establishing "quiet zones." To do so, communities must first mitigate the increased risk caused by the absence of a horn, such as implementing lights and gates at highway-rail grade crossings. 49 C.F.R. Part 222.

[41] A new rail POE is currently being studied in Santa Teresa, New Mexico, to divert rail traffic away from downtown El Paso, Texas. In addition, Laredo, Texas, has proposed three different locations for a new rail bridge over the years, although according to a representative from one railroad that operates through Laredo, none of the these proposals is currently being actively pursued.

[42] Pub. L. No. 111-5 123 Stat. 115 (2009). According to the county official, the costs for the bridge on the Mexico side were $80 million, for a total project cost of over $120 million.

[43] According to the Brownsville MPO representative, the City of Brownsville is responsible for zoning changes. This representative recommends changing the zoning in the immediate vicinity of the new rail corridor, which currently allows for residential development.

[44] GAO-14-740.

[45] We selected these communities based on BTS data on the number of inbound trains. BTS does not collect data on number of outbound trains or train length or speed.

[46] We received information from three railroads but information from one of these railroads was incomplete. We did not receive information from two railroads. As one railroad representative noted, it is problematic for railroads to obtain information on train speeds as speeds are typically managed by maintaining average speeds between points along a route's corridor. A representative from another railroad referred us to the national Highway-Rail Crossing Inventory for all data.

[47] This includes both inbound and outbound trains. While the data from these railroads allowed us to calculate examples of blockage times, they do not allow us to calculate the range of blockage times that might be experienced in communities with different rail patterns. In particular, if we had obtained data on trains with different lengths and different speeds, we may have identified a different range of blockage times.

[48] 49 C.F.R. Part 234, 80 Fed. Reg. 746 (Jan. 6, 2015). This final rule implemented section 204(a) of Rail Safety Improvement Act of 2008, Pub. L. No. 110-432, Division A, Title II (Oct. 16, 2008) codified at 49 U.S.C. § 20160.

[49] However, the Act stated that FRA shall develop a model of a state-specific highway-rail grade crossing action and distribute the plan to each state not later than one year after enactment. The plan shall include, among other things, methodologies for identifying and evaluating highway-rail grade crossing safety risks, including the risks posed by blocked highway-rail grade crossings due to idling trains. See Pub. L. No 114-94 § 11401 (2015).

[50] GAO-14-740.

End Notes for Appendix I

[1] These railroads were: Kansas City Southern Railway Company, Union Pacific Railroad Company, BNSF Railway Company, Canadian National Railway Company, and Canadian Pacific Railway.

[2] Brownsville was excluded because at the time of our visit in late June to early July 2015, the new international rail bridge was nearing completion, and as a result, the railroad was in the process of changing its travel pattern, making it difficult to characterize the impacts of freight rail on the community.

[3] Wc received information from 3 railroads but information from one of these railroads was incomplete.

In: Congestion Impacts of Freight Flows ISBN: 978-1-53610-033-4
Editor: Diana Hughes

Chapter 2

FREIGHT TRANSPORTATION: DEVELOPING NATIONAL STRATEGY WOULD BENEFIT FROM ADDED FOCUS ON COMMUNITY CONGESTION IMPACTS*

United States Government Accountability Office

WHY GAO DID THIS STUDY

Projected increases in the transport of freight by rail and truck may produce economic benefits but also increase traffic congestion in communities. MAP-21, which contains a number of provisions designed to enhance freight mobility, is currently before Congress for reauthorization. GAO was asked to review trends in freight flows and any related traffic-congestion impacts.

This report addresses among other things: (1) recent changes in U.S. rail and truck freight flows and the extent to which related traffic congestion is reported to impact communities, and (2) the extent to which DOT's efforts to implement MAP-21 address freight-related traffic congestion in communities. GAO analyzed rail data from 2007 through 2012 and highway data from 2010 and 2012 and reviewed 24 freight-related traffic congestion mitigation projects at 12 locations selected on the basis of different geographical locations and

* This is an edited, reformatted and augmented version of a United States Government Accountability Office publication, No. GAO-14-740, dated September 2014.

sizes. The results are not generalizable. GAO also reviewed federal laws and interviewed freight stakeholders.

What GAO Recommends

Congress should consider clarifying the purpose of the primary freight network and, as relevant to this purpose, revising the mileage limit requirement.

DOT should clarify the federal role for mitigating local freight-related congestion in the *National Freight Strategic Plan*, including a strategy for improving needed data. DOT concurred with the recommendations.

What GAO Found

Recent trends in freight flows, if they continue as expected, may exacerbate congestion issues in communities, particularly along certain corridors. As of 2012, the latest year for which data were available, national freight rail and truck traffic had approached levels of 2007 prior to the economic recession. Certain trends related to specific commodities have affected rail flows, including increases in domestic crude oil production. A key negative impact of increasing freight flows is congestion at highway-rail grade crossings, where road traffic must wait to cross the tracks when trains are passing. For example, a Miami-area study found that rail crossings in the area caused delays of roughly 235,000 person-hours per year at a cost of $2.4 million. Although several communities we visited had documented long-standing concerns over freight-related traffic congestion, state and local stakeholders we met with had varying levels of quantified information regarding the extent of the impacts or costs to the community. For example, in contrast to the Miami study, another study we reviewed included some information on train counts, but did not document hours of delay or any costs associated with such delays.

The Department of Transportation's (DOT) efforts to implement the freight-related provisions of the Moving Ahead for Progress in the 21st Century Act (MAP-21) are still underway but so far do not fully consider freight-related traffic congestion. MAP-21's freight policy goals do not explicitly include addressing freight-related traffic congestion, but MAP-21 requires

DOT to identify best practices to mitigate the impacts of freight movement on communities in a national freight strategic plan, which is due in October 2015. MAP-21's requirements and DOT's efforts so far do not fully establish the federal role or identify goals, objectives, or performance measures in this area, which may limit the usefulness of the *National Freight Strategic Plan*. For example:

- DOT issued for comment a required draft primary freight network, but according to DOT and other stakeholders, MAP-21's lack of defined purpose for the primary freight network and mileage limit of 27,000 miles hampered DOT's ability to include in this draft network some types of roads where local traffic congestion impacts of national freight movements are often experienced, such as roads connecting ports to freeways. The significance of the 27,000 mileage limitation is not clear. DOT released a surface transportation reauthorization proposal in April 2014 that proposed establishing a multimodal national freight network with a defined purpose and with no mileage limit.
- DOT is currently developing the *Freight Transportation Conditions and Performance Report*, which is to support the *National Freight Strategic Plan*. For this and other documents, DOT established a broad goal to reduce freight-related community impacts. However, DOT did not identify clear goals, objectives, or measures related to freight-related traffic congestion in local communities due to a lack of reliable national data. Thus, a clear federal role has not been established. High-quality data are essential to supporting sound planning and decision-making. Without reliable national data, it will be difficult for DOT to establish goals and objectives and to define the extent of freight-related traffic congestion and measure performance.

Abbreviations

AADT	average annual daily traffic
AAR	Association of American Railroads
ATA	American Trucking Associations
ATRI	American Transportation Research Institute
BNSF	Burlington-Northern Santa Fe
BTS	Bureau of Transportation Statistics

CMAQ	Congestion Mitigation and Air Quality Improvement
CREATE	Chicago Region Environmental and Transportation Efficiency Program
DOT	Department of Transportation
FHWA	Federal Highway Administration
FRA	Federal Railroad Administration
GPRA	Government Performance and Results Act
HPMS	Highway Performance Monitoring System
ITS	intelligent transportation systems
MAP-21	Moving Ahead for Progress in the 21st Century Act
MPO	metropolitan planning organization
NHS	National Highway System
NJ DOT	New Jersey DOT
SAFETEA-LU	Safe, Accountable, Flexible, Efficient Transportation Equity Act: A Legacy for Users
STB	Surface Transportation Board
TIFIA	Transportation Infrastructure Finance and Innovation Act
TIGER	Transportation Investment Generating Economic Recovery
UP	Union Pacific
WSDOT	Washington State DOT

* * *

September 19, 2014

The Honorable John Walsh
United States Senate

Dear Senator Walsh:

The nation's freight transportation network is vital to the functioning of the national economy. In 2012, the United States' transportation system moved 19.7 billion tons of goods, valued at more than $17 trillion, according to the Department of Transportation (DOT). Freight travels over an extensive network that consists of approximately 4 million miles of highways and roads and 140,000 miles of rail lines, as well as inland waterways, pipelines, and airways. Based on tonnage, trucks transport most freight in the United States (about 70 percent), but railroads also carry significant volumes over long distances. Freight trucks and trains both carry a variety of commodities. These include bulk goods—such as gravel, coal, and cereal grains—and consumer goods packed in containers as intermodal shipments that can be transferred among trains, trucks, and ships. DOT projects that the total tonnage of freight moved in the U.S. annually will increase 51 percent from 2007 to 2040 (from 18,879 million tons to 28,520 million tons).

While increased freight activity is closely associated with economic growth, this activity comes with a cost. Higher levels of freight movement can result in higher levels of congestion—both in the freight network itself[1] and in local communities where traffic on local highways and roads can be affected by truck or rail flows.[2] Freight-related traffic congestion may be due to highway-rail grade crossings, heavy freight traffic exiting and entering ports or other freight centers from local roads, or heavy truck traffic in rural communities related to energy production.[3] In recent years, some cities and community organizations in the upper Northwest have expressed concern that commodity trends—specifically, recent and potential increases in the transportation of crude oil from the Bakken region in North Dakota to the West Coast and the transportation of coal to proposed export terminals in Washington State—could dramatically increase these negative impacts.

In 2012, the President signed into law the Moving Ahead for Progress in the 21st Century Act (MAP-21), which authorized most surface transportation programs for 2 years—the first surface transportation authorization since 2005. MAP-21, which expires on September 30, 2014, contains provisions designed to enhance freight movement in support of national goals. In deliberations in advance of reauthorizing this legislation, Congress may consider additional provisions related to freight transportation.

You asked us to review how trends in particular commodity flows are affecting local freight-related traffic congestion and how communities may mitigate such impacts.[4] Specifically, this report addresses: (1) how U.S. rail and truck freight flows are changing, and the extent to which freight-related traffic congestion is reported to impact local communities; (2) how communities have funded efforts to mitigate freight-related traffic congestion, and what funding challenges, if any, communities report facing; and (3) the extent to which DOT's efforts to implement freight-related provisions of MAP-21 have addressed local freight-related traffic congestion.

To address these objectives, we analyzed Surface Transportation Board (STB) data on rail carloads from 2007 to 2012. We chose to analyze STB data beginning in 2007 after being advised by STB officials that to examine recent trends, 2007 was more indicative than 2008 of rail traffic before the recent U.S. economic recession. We also analyzed DOT data on 2012 annual average daily traffic of vehicles and on 2010 highway freight flows, and the Federal Railroad Administration's (FRA) National Grade Crossing Inventory and FRA's flows over the rail network of STB's rail data from 2010 and 2012. We used the most recent data available for the purposes of our analyses and concluded the data were sufficiently reliable for the purposes used in our report. We reviewed the reliability of these data by reviewing related documentation and interviewing agency officials about their data collection procedures, among other things. We also obtained information from the Association of American Railroads (AAR) on rail carload levels in 2013.

After learning that there was not a federal source of data that could reliably be used to analyze freight truck trends from 2007 to 2012, because, among other things, the data do not sufficiently distinguish among classes of trucks, we reported information available in studies by the American Trucking Associations (ATA) and the American Transportation Research Institute (ATRI), the research arm of ATA, which performs some of its studies in partnership with the Federal Highway Administration (FHWA). We reviewed DOT's evaluation of the ATA truck tonnage data and the methodologies used in ATRI studies and determined that they were sufficiently reliable to use to provide contextual information. We also obtained some information on freight truck levels in 2013 from ATA and ATRI.

We reviewed freight-related traffic congestion issues and mitigation projects at 12 selected communities to understand the types of local impacts communities are experiencing from freight movements and efforts they are making to mitigate such impacts. We selected communities representing a variety of different geographical locations and sizes, that had either successfully developed projects to mitigate freight-related congestion impacts (such as railroad-highway grade separations or truck bypass routes), were on record as having studied and attempted to mitigate such impacts, or were identified by DOT officials as having recently had an increase in freight-related traffic congestion. Selected communities included Seattle, Edmonds, and Spokane, Washington; Billings and Glendive, Montana; Williston (and surrounding communities in the Bakken oil field region), North Dakota; Chicago, Illinois; Gulfport and Pascagoula, Mississippi; Miami, Florida; Paulsboro, New Jersey, and the New York City metropolitan region, including projects for New Jersey and New York. Within these communities, we reviewed 24 ongoing or recently completed congestion mitigation projects, such as grade separations, port rail improvements and access roads, and truck bypass routes. Findings from our selected communities and projects are not generalizable.

We also reviewed and analyzed policy documents, proposed legislation, and federal laws, including MAP-21, DOT's documentation of efforts to implement MAP-21's freight provisions, such as its published draft primary freight network, and public comments on the draft primary freight network, as well as DOT's April 2014 surface transportation reauthorization proposal. Finally, to inform all objectives, we reviewed prior GAO work on surface transportation and interviewed DOT, state, local, and private industry officials involved in freight issues. These included officials from DOT's FRA, Federal Highway Administration (FHWA), and Bureau of Transportation Statistics; and STB; state and local transportation agencies in the communities we visited, U.S. railroad companies including the Burlington-Northern Santa Fe (BNSF), Union Pacific (UP), CSX, and Norfolk Southern Corporation; and industry associations such as AAR, ATA, ATRI, the Owner Operator Independent Drivers' Association, and the American Association of State Highway and Transportation Officials.

We conducted this performance audit from September 2013 to September 2014 in accordance with generally accepted government auditing standards. Those standards require that we plan and perform the audit to obtain sufficient, appropriate evidence to provide a reasonable basis for our findings and conclusions based on our audit objectives. We believe that the evidence obtained provides a reasonable basis for our findings and conclusions based on our audit objectives. For more information on our scope and methodology, see appendix I.

Background

Our nation's freight is moved by different transportation modes—including rail, truck, ships, pipelines, and airplanes—with the majority of freight by weight reaching its final destination by either truck or by rail. In addition, some freight movements are intermodal, such as containers that arrive in the U.S. by ship and are subsequently transferred to rail or truck. A map of 2010 freight flow volumes across the U.S. shows that many states had at least 100 million tons of freight annually flowing over some highways, railways, or inland waterways (see fig. 1).[5]

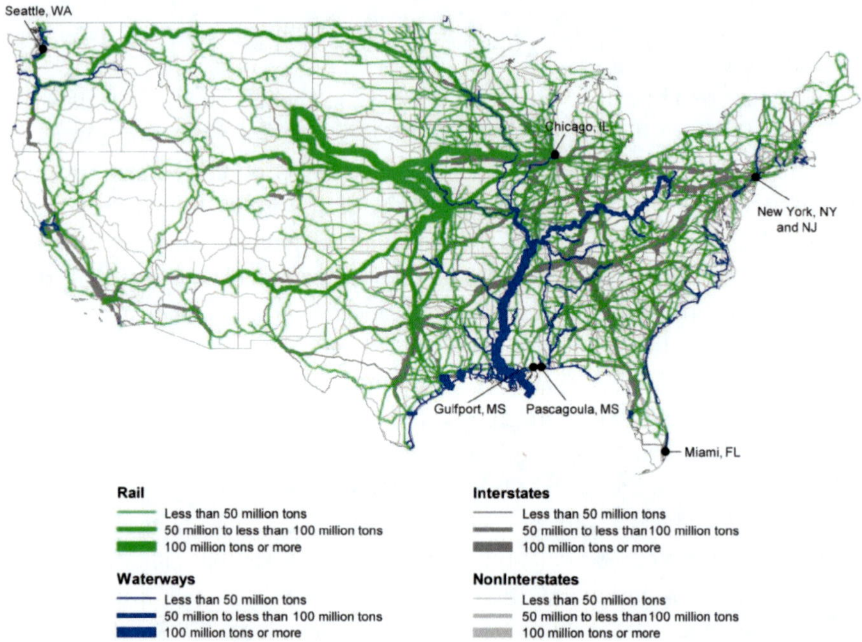

Source: GAO presentation of U.S. Department of Transportation and Surface Transportation Board information. | GAO-14-740.

Figure 1. Freight Flows, by Tons, Moved on Railways, Highways, and Inland Waterways, 2010, with Ports Serving Selected Communities.

Recent commodity trends, such as the increase in domestic crude oil production from shale formations, have affected freight movements. [6] According to the Energy Information Administration, oil production from shale formations, such as North Dakota's Bakken region, has increased 6 fold between 2007 and 2012, and is expected to continue to increase—by 48 percent from 2012 to 2019—and remain above 2012 levels through 2040.

While vital to our nation's economy, freight movements can contribute to two types of congestion impacts, which can be temporary or more permanent in nature. The first is congestion on the freight network itself, such as when congested rail corridors and highways result in delays for freight shippers. The second, which is the focus of this report, is freight-related traffic congestion in communities. Unlike congestion on the freight network, freight-related traffic congestion may not delay or affect freight shippers but results in delays and congested road conditions for passenger and emergency response vehicles. For example, such vehicles may experience delays at highway-rail grade crossings or near rail yards and transfer facilities. In addition, delays and congested road conditions can occur when freight trucks must traverse local streets to access major freight centers, such as ports or distribution warehouses.

Depending on the specific circumstances in a community, various types of projects can be undertaken to help alleviate freight-related traffic congestion (see fig. 2). For instance, grade separation projects—such as overpasses and underpasses—can eliminate congestion issues associated with heavily traveled highway-rail grade crossings. In addition, signalization projects, including the use of intelligent transportation systems (ITS), can help reduce delays at highway-rail grade crossings by providing drivers with advanced warning of approaching trains or extending green lights after a train has cleared an intersection.[7] To reduce truck traffic on local roads, communities may build direct freeway connections between freight centers and nearby highways (connector roads or tunnels), redirect freight trucks around towns (bypass routes), or improve the ability of local roads to manage truck traffic—for example, by widening lanes and adding passing lanes and traffic lights.[8]

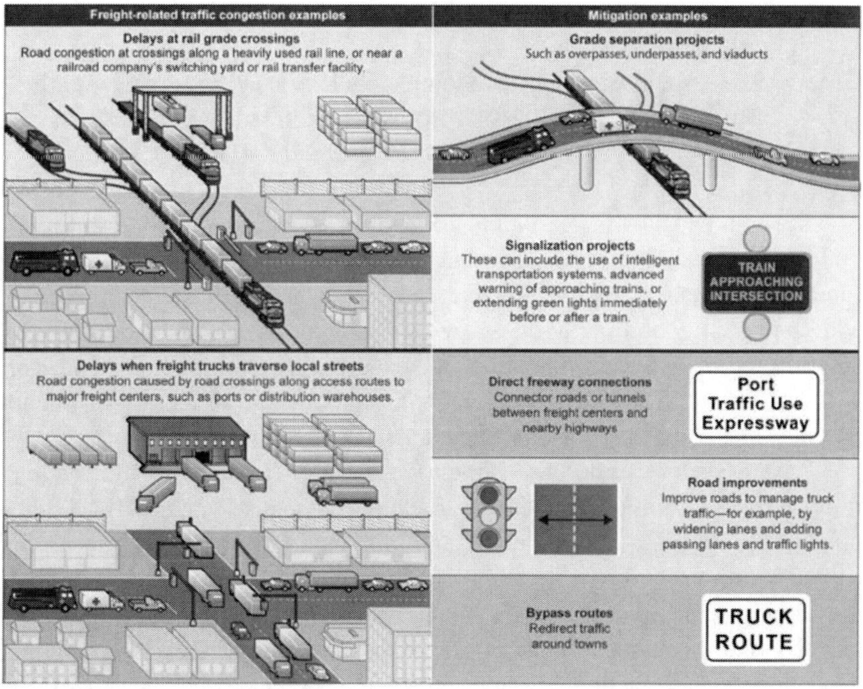

Source: GAO. | GAO-14-740.

Figure 2. Examples of Freight-Related Traffic Congestion and Mitigation Efforts.

Multiple stakeholders play a role in mitigating the impacts of freight-related traffic congestion in communities. At the federal level, DOT and its component agencies develop and enforce regulations, advance national strategic goals by providing technical assistance and leadership through collaboration with various transportation stakeholders, and oversee federal-funding programs that can be used to mitigate freight-related traffic congestion (see table 1 below for brief descriptions of selected programs). Many of these funding programs had been available for freight-related congestion-mitigation and traffic-flow improvement projects prior to MAP-21.

- *FHWA* distributes about $40 billion a year to states to build, preserve, and improve the nation's road and highway infrastructure through multiple formula and discretionary grant programs, collectively known as the federal-aid highway program.[9]
- *FRA* enforces safety regulations, makes selective investments in the country's rail network, and maintains the National Grade Crossing Inventory, a uniform, national database meant to assist in the planning and implementation of rail-highway crossing safety improvement programs.
- *STB*—an independent economic regulatory agency administratively affiliated with DOT—issues licenses to railroads for construction and acquisition projects and adjudicates railroad rate and service disputes.

State and local transportation agencies have primary responsibility for building, maintaining, and operating roads and planning and prioritizing funding for transportation projects, including freight-related projects. In addition, public port authorities—which may be local, regional, or state authorities—oversee port operations, manage port revenue streams, and work with state and local transportation agencies to plan freight infrastructure projects, including projects to mitigate traffic congestion impacts of freight.[10]

Private sector entities are largely responsible for moving the nation's freight. These entities—including trucking companies, railroads, airlines, pipeline companies, and terminal and distribution-center operators—often compete with each other for freight-shipping business and make key routing, operating, and equipment investment decisions. Some private entities, such as railroad and pipeline companies, own the freight infrastructure they are dependent upon to facilitate freight movements; other private entities, such as trucking firms and terminal operators, rely on what is primarily publicly-owned freight infrastructure, including the National Highway System.

Table 1. Federal Transportation Funding and Financing Programs That Can Be Used to Mitigate Freight-Related Traffic Congestion

Program	Lead administration or office within DOT	Purpose
Railway-Highway Crossings Program[a]	Federal Highway Administration	To fund safety improvements to reduce the number of fatalities, injuries, and crashes at public grade crossings.
Congestion Mitigation and Air Quality Improvement Program	Federal Highway Administration	To fund transportation projects or programs that will contribute to attainment or maintenance of the National Ambient Air Quality Standards for ozone, carbon monoxide and particulate matter by reducing congestion and improving air quality.
National Highway Performance Program	Federal Highway Administration	To support the condition and performance of the National Highway System (NHS), construct improvements to the NHS, and ensure that federal highway construction investments help states achieve the performance targets outlined in their NHS asset management plans.
Highway Safety Improvement Program	Federal Highway Administration	To achieve a significant reduction in traffic fatalities and serious injuries on all public roads, including non-State-owned public roads and roads on tribal lands.
Surface Transportation Program	Federal Highway Administration	To preserve and improve the condition and performance of any federal-aid highway, bridge, or tunnel project on any public road; pedestrian and bicycle infrastructure; and transit capital projects.
Transportation Infrastructure Finance and Innovation Act Credit Assistance Program	Office of the Secretary	To fill market gaps and leverage substantial private co-investment by providing federal credit assistance— in the form of direct loans, loan guarantees, and lines of credit—to eligible surface

Table 1. (Continued)

Program	Lead administration or office within DOT	Purpose
		transportation projects, including highway, transit, intercity passenger rail, some types of freight rail, and intermodal freight transfer facilities.
Transportation Investment Generating Economic Recovery Grant Program	Office of the Secretary	To fund projects (via a competitive grant program) that have a significant impact on the Nation, a region or a metropolitan area in terms of safety, economic competitiveness, state of good repair, livability and environmental sustainability.
Railroad Rehabilitation and Improvement Financing Program	Federal Railroad Administration[b]	To finance the development of railroad infrastructure—such as track, bridge, and rail yard improvements—through direct loans andloan guarantees.

Source: GAO | GAO-14-740.

[a]States' Railway-Highway Crossings Program funds are set aside from their Highway Safety Improvement Program apportionment.

[b]Another Federal Railroad Administration program—the Rail Line Relocation and Improvement Capital Grant Program—is intended to assist state and local governments in mitigating the adverse effects created by the presence of rail infrastructure; however Congress has not appropriated funding for this program since fiscal year 2011.

To mitigate freight-related traffic congestion and meet other transportation infrastructure needs, public and private transportation stakeholders may enter into what are referred to as public-private partnerships. As we have previously reported, these partnerships can offer benefits to the public sector, such as the sharing or transferring of projects' financial risk to the private sector and increased operational and management efficiency of publicly-owned infrastructure, depending on the specific partnership agreement.[11] In more straightforward examples of public-private partnerships, private sector entities may provide direct funding to publicly-sponsored infrastructure projects, or enter into contracts with public agencies to design and construct these projects. Other partnerships can be more complex. For example, a private entity may assume control over the operations and maintenance of an existing public asset, such as a toll road, for a fixed period of time as part of a revenue-sharing agreement with the public sector. Alternatively, a private entity may take responsibility for all aspects of a new, publicly-sponsored infrastructure project—including design, finance, construction, and operations and maintenance—in exchange for periodic, performance-based payments from the public sector, from which the private sector expects to profit. In prior work, we have noted that there is no "free" money in public-private partnerships—for example, while highway public-private partnerships can be used to obtain financing for highway infrastructure without the use of public sector funding, this funding is a form of privately issued debt that must be repaid to private investors seeking a return on their investment by collecting toll revenues.[12]

In recent years, the federal government has placed an increasing emphasis on the nation's freight transportation system in its surface transportation authorizing legislation. In 2012, the President signed into law the Moving Ahead for Progress in the 21st Century Act (MAP-21)[13]. MAP-21 established seven goals for a national freight policy.[14] None of these goals explicitly encompasses addressing community impacts of freight, such as local traffic-congestion impacts. One of the goals is to invest in infrastructure improvements and implement operational improvements that reduce congestion—which could potentially incorporate local traffic congestion caused by freight flows. However, according to DOT officials, DOT has generally treated congestion as a source of delay in the delivery of freight shipments and hence as a factor that reduces economic productivity, not as a factor that produces adverse community impacts. In addition, another goal, to reduce the environmental impacts of freight movement on the national freight network, refers to a type of community impact that can be caused by freight congestion. Unlike its predecessor legislation[15], MAP-21 also requires DOT to establish a national freight strategic plan in consultation with state departments of transportation and other transportation stakeholders. MAP-21 includes as a required element of the freight strategic plan the identification of best practices to mitigate the impacts of freight movement on communities.

Also unlike its predecessor legislation, MAP-21 requires DOT to establish a national freight network. The national freight network is to include a primary freight network and critical rural freight corridors to assist states in strategically directing resources toward improved system performance for efficient movement of freight on highways. The act established a deadline of October 1, 2013 for DOT to designate the highway primary freight network. In

November 2013, DOT released for comment a draft primary freight network.[16] MAP-21 also requires DOT to develop a freight transportation conditions and performance report by October 1, 2014, which is to feed into the development of a national freight strategic plan by October 1, 2015. Moreover, MAP-21 requires DOT to encourage each state to develop a freight advisory committee and a state freight plan. MAP-21 directly addresses our prior suggestion that to maximize the efficient movement of freight, Congress should consider defining the federal role in surface transportation.[17]

In March 2014, we found that DOT had partially addressed our 2008 recommendation to develop a comprehensive national strategy related to freight, citing DOT's progress in (1) developing planning tools, transportation investment data, and performance measures for planning and evaluating freight projects; (2) completing a draft of the highway primary freight network; and (3) establishing a National Freight Advisory Committee. [18] The National Freight Advisory Committee—comprised of public and private-sector freight stakeholders appointed by the Secretary of Transportation—provides advice and recommendations to DOT on matters relating to freight transportation in the United States and the implementation of the freight provisions of MAP-21. In June 2014, the National Freight Advisory Committee issued its final recommendations to DOT for consideration in development of a national freight strategic plan.

In April 2014, DOT sent to Congress a multi-year surface transportation proposal, called Generating Renewal, Opportunity, and Work with Accelerated Mobility, Efficiency, and Rebuilding of Infrastructure and Communities throughout America (GROW America). This proposal was introduced in the House of Representatives on June 11, 2014, and referred to appropriate committees.[19] According to DOT officials, this proposal demonstrates further progress in addressing our 2008 recommendation that DOT develop a comprehensive national strategy related to freight, by proposing, among other things, $10 billion in funding for a dedicated freight program, a multimodal approach to freight funding, requirements for states to develop state freight plans and to create state freight advisory committees, incentives for states to engage in multi-state collaborative freight planning, use of benefit-cost analysis in the evaluation of freight projects, and use of retrospective analysis to evaluate the effects of freight projects after they have been completed. The proposal would also establish a National Freight Infrastructure Program that would have, among other goals, an explicit goal to reduce adverse community impacts of freight transportation.

INCREASING FREIGHT FLOWS MAY EXACERBATE LONG-STANDING COMMUNITY CONGESTION ISSUES

Freight rail traffic has increased since 2009, and as of 2013, the total number of rail carloads approached levels prior to the 2008 recession. State and local officials in the upper Northwest have raised concerns regarding rapid increases in carloads of crude oil transported by rail, largely from North Dakota, and the increase in carloads of coal being transported by rail to ports. Traffic congestion at highway-rail crossings— a long-standing concern in many local communities—may be exacerbated by increased rail freight flows; however, the extent of the impacts and costs are often not well documented. Truck volumes have also increased in recent years and are approaching levels prior to the 2008 recession, causing community impacts when trucks must travel on local roads, such as roads connecting ports to freeways. Moreover, in recent years, certain areas, such as the Bakken region of North Dakota, have had large increases in truck traffic on local roads due to increasing oil extraction.

Freight Rail Traffic Has Risen Recently and Increased Carloads of Certain Commodities Have Raised Concerns

Rail Traffic Increased since 2008 Downturn, with Commodity Trends Affecting Rail Traffic in Certain States

According to STB data, nationally, since 2000, freight rail traffic reached a high in 2006, dropped slightly in 2007, and then fell sharply with the economic recession of 2008 and 2009 before rising again starting in 2010. Our analysis of STB rail data from 2007 through 2012[20] showed that the total number of carloads in 2012 (about 34 million) remained 8 percent below the total number of carloads in 2007 (about 37 million), although it represented a significant increase compared to the low of 2009 (about 30 million). Although 2013 STB data were not available for this report, according to AAR officials, AAR data show that the total number of carloads in 2013 continued to climb toward 2007 levels.

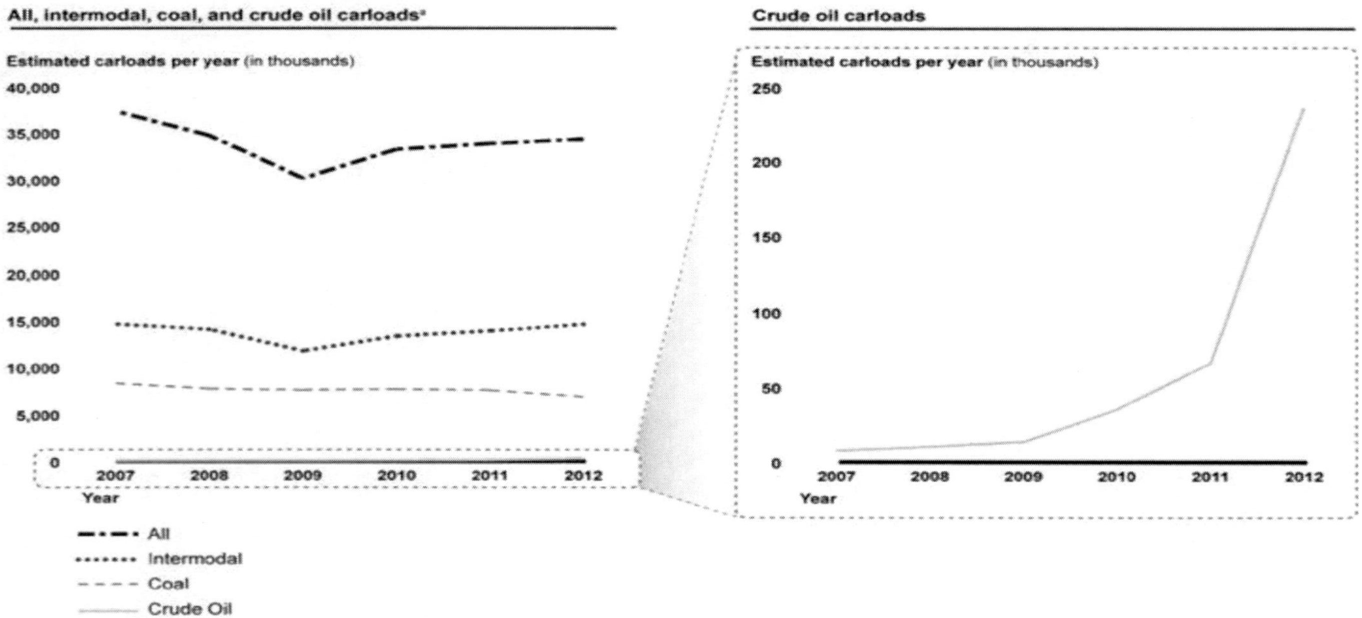

Source: GAO analysis of Surface Transportation Board Data. | GAO-14-740.

[a]For all, intermodal, and coal carloads, the margin of error as a percentage of the total is less than 5 percent. Crude oil carloads, which existed in small numbers in 2007 through 2011, had for those years a declining margin of error that started at 26 percent in 2007 and went down to about 8 percent in 2011. For 2012, the margin of error for crude oil carloads was less than 5 percent.

Figure 3. Estimated Rail Carloads of Selected Shipment Types, and of Crude Oil Separately 2007-2012.

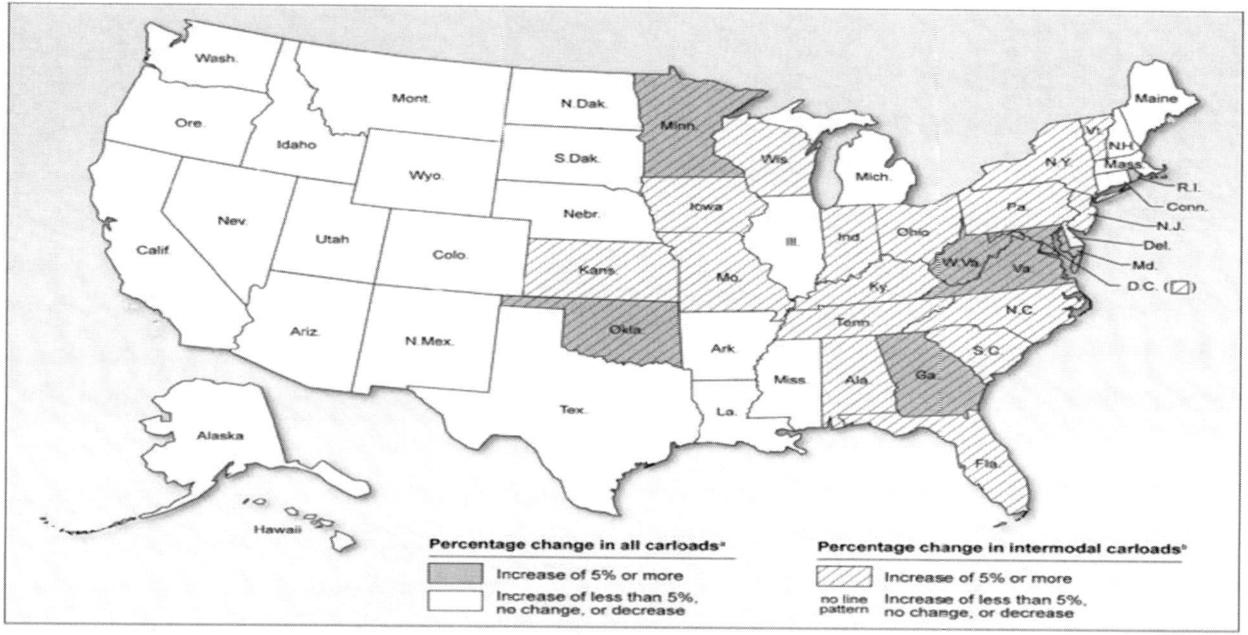

Source: GAO analysis of Surface Transportation Board Data. | GAO-14-740.

[a]Maryland had the largest percentage increase in all carloads, of about 60 percent. The increases in all carloads in states with an increase of 5 percent or more are statistically significant with the exception of Rhode Island.

[b]Maryland also had the largest percentage increase in intermodal carloads, of about 180 percent. The increases in intermodal carloads in states with an increase of 5 percent or more are all statistically significant.

Figure 4. Estimated Change in Total Carloads and Intermodal Carloads, by State, 2007–2012.

While according to AAR officials, the recession and subsequent recovery were the major factors affecting total carload trends during this period, several trends related to shipment type or specific commodities have also affected rail flows since 2007. These trends include an increase in intermodal traffic, a decrease in overall coal traffic but an increase in coal traffic related to export, and a sharp increase in crude oil transported by rail. Although crude oil represented about 1 percent of all rail carloads in 2012,[21] the percentage increase in crude oil carloads from 2007 to 2012—of over 3,000 percent, from about 7,000 carloads to about 236,000 carloads—was far higher than the percentage change in carloads of any other commodity.[22] Figure 3 shows the trends since 2007 for all carloads and for intermodal, coal, and crude oil carloads.

Recent trends related to intermodal, coal, and crude oil shipped by rail affected rail traffic in different states differently. In the context of an 8 percent decline in the total number of carloads nationwide from 2007 through 2012, only 6 states, mostly in the Eastern half of the country, experienced a statistically-significant increase of at least 5 percent in the total number of rail carloads during this period;[23] all of these states also had increases in the number of intermodal carloads (see fig. 4).[24] Intermodal traffic, which represented about 43 percent of all rail carloads in 2012, was described by DOT and AAR officials as an area of strong growth in the rail industry. According to DOT officials, two major corridor projects initiated by Norfolk Southern railroad and funded in part by the federal government may have facilitated the growth of intermodal rail traffic in the Eastern part of the U.S. in 2012 and 2013 and may continue to do so in future years. The Crescent Corridor is designed to facilitate the movement of intermodal traffic by rail from the Gulf Coast to the Northeast and was the recipient of a 2010 Transportation Investment Generating Economic Recovery (TIGER) grant of $105 million as well as Congestion Mitigation and Air Quality Improvement (CMAQ) funds. The Heartland Corridor was designed to permit double stack container trains to travel directly from the Port of Virginia to Columbus and Chicago. This project received an authorization of federal funds of about $126 million from the Safe, Accountable, Flexible, Efficient Transportation Equity Act: A Legacy for Users (SAFETEA-LU) as well as federal funds from the American Recovery and Reinvestment Act of 2009 and a 2011 TIGER grant of $12 million.[25] Both projects have a goal of reducing freight-related truck congestion.

Like the increase in intermodal rail traffic, increases in crude-oil and coal carloads carried by rail for export predominantly occurred in certain states.

Figure 5 shows the states that had an increase of 10,000 carloads or more of coal or oil from 2007 through 2012. According to the Energy Information Administration, the increase in crude-oil carried by rail was mostly due to the production of oil from shale areas such as the Bakken centered in North Dakota. Seventy five percent of crude oil rail carloads that originated in the U.S. in 2012 originated in North Dakota.[26] One reason for the increase in crude oil being shipped by rail is the limitation of the nation's pipeline capacity to handle current oil production. In March 2014, we found that most of the system of crude oil pipelines in the United States was designed primarily to move crude oil from the South to the North; emerging crude oil production centers in Western Canada, Texas, and North Dakota have strained the existing pipeline infrastructure, and in some areas pipeline capacity has been inadequate.[27] For example, according to the North Dakota Department of Transportation's State Freight Plan, North Dakota has lacked adequate pipeline capacity to handle the huge increase in crude oil production since 2008—from about 122,000 barrels a day in 2008 to about 933,000 barrels per day in January 2014—with the state's crude-oil pipeline capacity in 2013 being 583,000 barrels per day. According to BNSF officials, pipelines in these areas are at or near capacity, creating a strong demand for additional outbound transportation options, as well as a demand for inbound oil field supplies. Most crude oil transported by rail originated in North Dakota and, according to AAR, largely went to refineries in the Gulf Coast, the East Coast, and the West Coast, passing through other states along the way (see fig. 5). According to AAR officials, these states are likely to have had continued increases in the amount of crude oil travelling by rail in them since 2012, with AAR reporting that Class I freight railroads originated 407,761 carloads of crude oil in 2013.[28]

With the overall decrease in transportation of coal by rail, only 5 states experienced an increase of at least 10,000 carloads of coal from 2007 to 2012. All of these states have ports from which coal is shipped for export or are on the way to such ports from Wyoming or West Virginia, where the majority of U.S. coal being transported by rail originates. According to several industry officials we spoke with, in many of these states, the amount of crude oil and coal being shipped by rail is likely to continue to increase in coming years. According to the Energy Information Administration, crude oil production is expected to continue to increase in the near future,[29] while coal exports are expected to remain steady through 2020 and then increase through 2040.

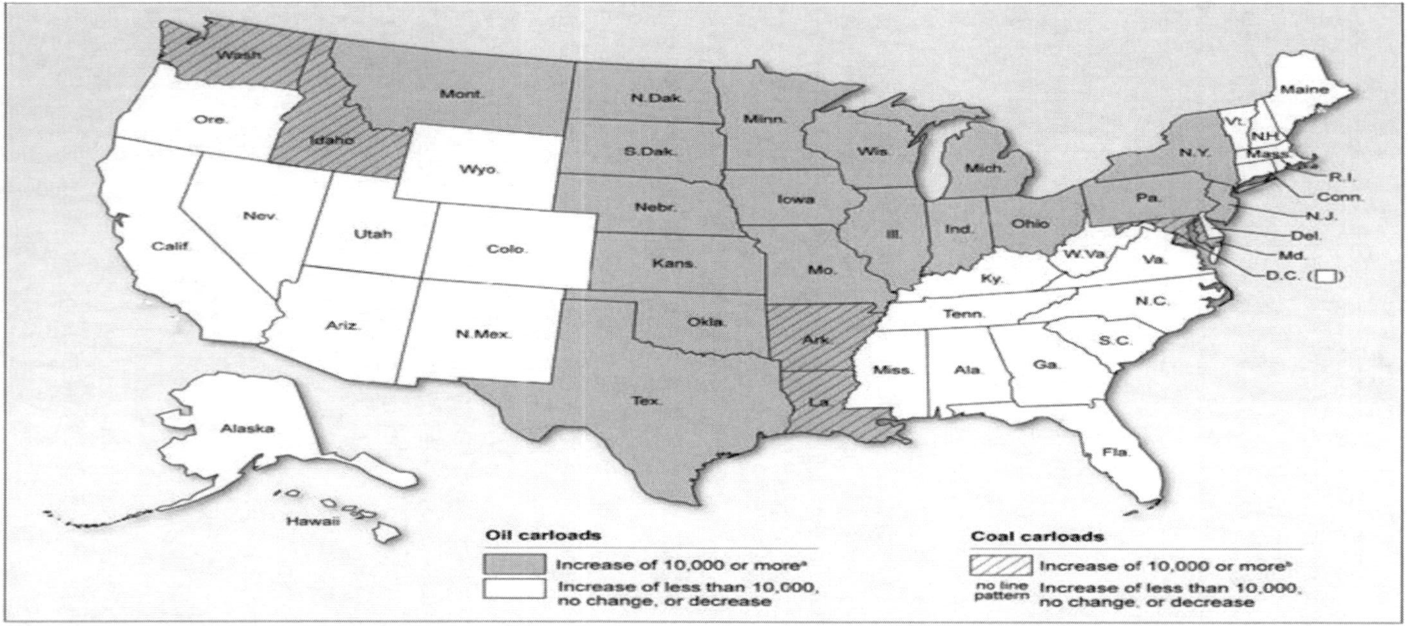

Source: GAO analysis of Surface Transportation Board Data. | GAO-14-740.

[a]North Dakota had the largest increase in the number of crude oil carloads, about 172,000. The increases in crude oil carloads in states with an increase of 10,000 carloads or more are all statistically significant. For this figure we used increase in number of carloads rather than percentage increase because in 2007, most states had no carloads of crude oil in them.

[b]Louisiana had the largest increase in the number of coal carloads, about 72,000. The increases in coal carloads in states with an increase of 10,000 carloads or more are all statistically significant with the exception of Arkansas.

Figure 5. Estimated Change in Carloads of Crude Oil and Coal, by State 2007–2012.

North Dakota officials in the Bakken region told us that they expect crude oil extraction in the region to increase for at least the next 10 years and production to continue to be higher than current rates for at least 20 years, with some production continuing to occur over the next 50 years or beyond. There is the potential that pipelines could eventually be built to carry much of this oil. However, according to AAR, rail offers greater flexibility than pipelines to shift product quickly in response to market needs and price opportunities. As a result, AAR officials predicted continued growth in the amount of crude oil shipped by rail in coming years.[30] With regard to coal traffic, the issuance of some proposals for new facilities that can handle coal exports in Washington state and Western Canada suggest that the industry expects continued growth in coal exports. If these terminals are developed and such growth occurs, it may lead to increases in the amount of coal being carried by rail to Washington State and Western Canada.

Local Concerns about Increased Rail Traffic Are Focused on Certain Commodities

Many state and local officials we spoke with from Washington, Idaho, and Montana expressed concern over impacts to local communities due to continued expected increases in crude oil and coal carried by rail to west coast refineries or terminals. Our analysis of 2012 data on rail carloads did not show an increase in the total number of all carloads in these states as of 2012. However, as shown above, our analysis did show increases in carloads of crude oil in all three of these states and of carloads of coal in Washington and Idaho. Much coal and oil transported by rail is carried on unit trains, which are trains that carry only one commodity, and are readily identifiable. In addition, unit trains can be long—for example, crude oil unit trains may consist of more than 100 cars, which could create noticeable waits at highway-rail grade crossings in local communities—although trains carrying other types of goods, such as intermodal trains, may also have more than 100 cars. Moreover, it is possible that these states will experience increased train traffic due to trends in these commodities in coming years. One issue affecting the level of concern among some state and local officials is that much of the increase in coal and crude-oil rail traffic in some states is passing through the state on its way from and to other locations. When, for example, oil is produced in a state or coal is delivered to a state's port for export, the state may experience negative impacts but may also reap economic benefits, such as increased jobs and tax revenue. However, according to some state and local officials we spoke with, when an energy commodity passes through a state on a rail car, the state gets no direct economic benefit from that commodity but may be incurring congestion-related and other impacts at highway-rail grade crossings. Figure 6 shows the states that had increases of over 10,000 annual carloads of crude oil or coal passing through the state from 2007 to 2012.

Traffic Congestion at Highway-Rail Crossings is a Longstanding Concern but Impacts and Costs Are Not Always Documented

DOT, state, and local officials identified a key negative impact of increasing freight flows as freight-related traffic congestion occurring at highway-rail grade crossings, where road traffic must wait to cross the tracks when trains are passing. Highway-rail grade crossings exist in every state in the U.S. In 2012, there were 210,621 public and private rail crossings in the U.S.[31] Figure 7 shows rail crossings that in 2012 included the passage of at least 200,000 rail cars annually and at least 10,000 daily vehicles[32] (see fig. 7).

Several communities we visited had documented longstanding concerns over highway-rail grade crossings.[33] For example, a 2004 study of potential grade separation projects in Billings, Montana, described seven previous studies of the same issues, with the earliest study from 1958. In the Seattle area, a corridor effort to reduce community impacts of freight through 20 grade separation projects, among other things, has been ongoing since 1998. Another corridor project—the Bridging the Valley Project—which is aimed at eliminating 75 highway-rail crossings along a 42 mile corridor between Spokane, Washington, and Athol, Idaho, was officially started in 2000, but according to Spokane officials, was designed to address at-grade crossing issues identified as problematic in the 1990s. And in Chicago, the Chicago Region Environmental and Transportation Efficiency Program (CREATE), which includes 25 grade separation projects as part of a larger number of projects designed to improve the fluidity of freight and passenger rail service throughout the region, was announced in 2003.[34] State and local officials described the following types of negative community impacts of traffic being blocked at highway-rail grade crossings:

- *Delays to motorists.* Such delays can be significant, such as when a large number of vehicles on the road results in a queue that can take time to disperse even after a continuously moving train has departed, or when a stopped train blocks the tracks for significant periods of time.

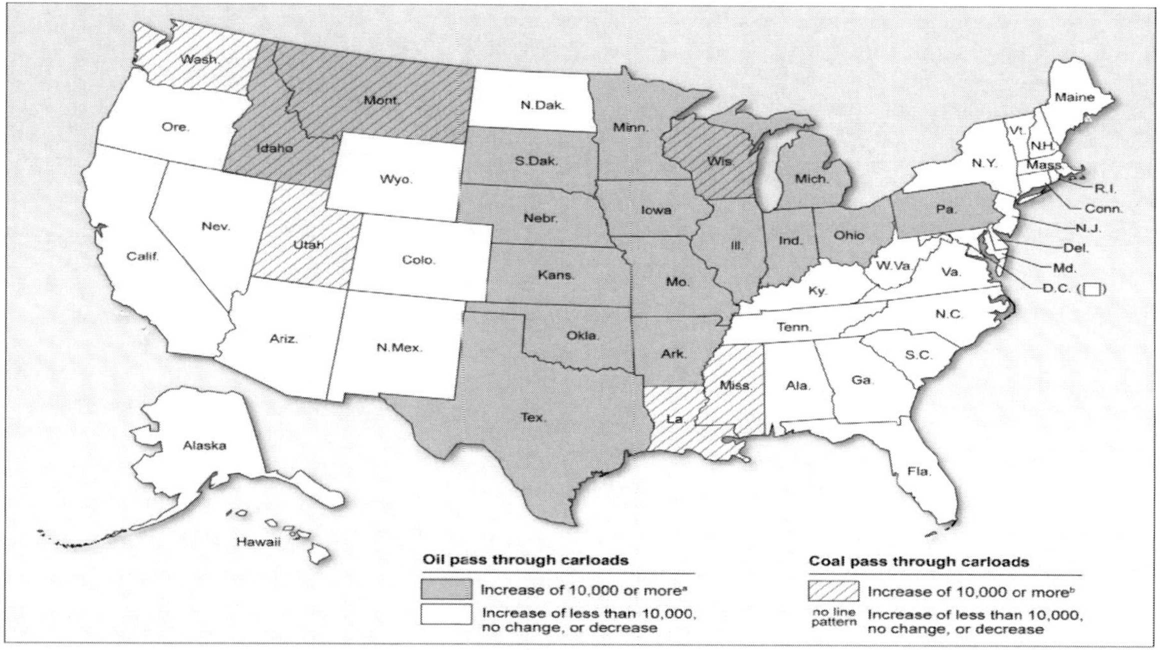

Source: GAO analysis of Surface Transportation Board Data. | GAO-14-740.

[a]Minnesota had the largest increase in the number of pass-through carloads of crude oil, about 165,000. The increases in crude oil pass through carloads in states with an increase of 10,000 carloads or more are all statistically significant. For this figure we used increase in number of carloads rather than percentage increase because in 2007, most states had no carloads of crude oil passing through.

[b]Washington State had the largest increase in the number of pass-through carloads of coal, about 75,000. The increases in coal pass through carloads in states with an increase of 10,000 carloads or more are all statistically significant.

Figure 6. Estimated Change in Pass-Through Carloads of Crude Oil and Coal, by State, 2007-2012.

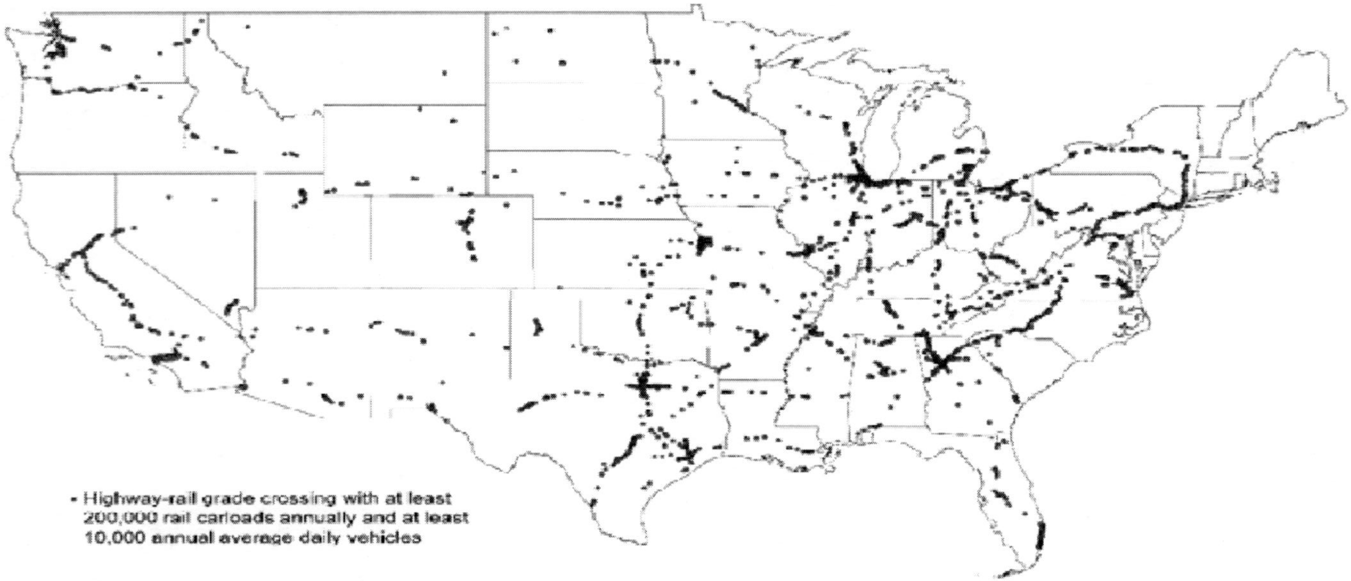

Source: GAO analysis of DOT and STB data. | GAO-14-740.

Note: In order to illustrate the location of highway-rail grade crossings that have above a certain threshold of vehicular and rail traffic, we used the most recent, reliable data available for both types of traffic, even though vehicular traffic was recorded in terms of annual average daily traffic, while rail carload information was recorded in terms of annual carloads.

Figure 7. Highway-Rail Grade Crossings with a Crossing of at Least 200,000 Rail Cars Annually and 10,000 Vehicles Daily, 2012.

- *Blocked emergency vehicles.* According to local officials, in Baker, Montana, a fire recently destroyed a house because train traffic had blocked the only two crossings in the town and prevented fire crews from responding in time. According to STB officials, stakeholders also often worry that ambulances will be unable to reach them or transport them quickly to hospitals because of trains blocking crossings.
- *Quality of life impacts.* Such impacts can be exacerbated by the specific location of the crossing in relationship to the community. For example, in Edmonds, Washington, a main rail line crosses the only access road to the town's ferry, which carried over 22.5 million riders in 2013. According to city officials, about 40 freight or commuter trains pass this crossing each day and can cause vehicle queues that extend 1.5 miles from the ferry terminal, which can be especially problematic because people are trying to make their scheduled ferries. In Billings, Montana, the community's long-standing concerns center on a highway-rail grade crossing in which a main rail line bisects one of the city's central downtown streets (see fig. 8). In addition to concerns related to delay and blocked motorists, several stakeholders described frustration that the community is divided when the crossing is blocked.

Source: GAO. | GAO-14-740.

Figure 8. A Highway-Rail Grade Crossing in Billings, Montana, Bisects the City.

In spite of often long-standing concerns with the negative effects of highway-rail grade crossings, communities we visited had varying levels of quantified information on impacts such as traffic delay times or costs. For example, the 2004 Billings, Montana, study described above included some general information on train counts but no information documenting the number of vehicles delayed, the hours of delay, or any costs associated with such delays. The CREATE program in Chicago has some information in its studies on estimated hours of delay and, according to a Illinois Department of Transportation official, used DOT's guidance to assign a value of time to provide an economic benefits calculation in its TIGER applications. According to Illinois Department of Transportation officials, the CREATE partners also use a model for various analyses that provides information on train numbers, length, weight, and speed.

However, according to Illinois state officials, efforts to quantify impacts of highway-rail grade crossings both for the CREATE program and for the Chicago area generally have been limited by a lack of current information on gate down time, as well as by a lack of specific information on the types, timing, and occupancy levels of vehicle traffic at the crossings. Moreover, according to an Illinois Commerce Commission official, efforts to accurately quantify the impact of highway-rail grade crossings at a high level—in Illinois or nationally—have been limited by a lack of data available on the number of trains and train lengths assigned by date, speed, and time. On the other hand, a study of delays caused by highway-rail grade crossings in the Miami area did take some of these factors into account. This study concluded that rail crossings in the area caused delays of roughly 235,000 person-hours per year at a cost of $2.4 million.[35]

Similarly, some efforts to document impacts of highway-rail grade crossings in Seattle do include detailed information on such things as train numbers, timing, and speed, in some cases provided by railroad companies, as well as detailed traffic information.

Among other issues, DOT officials stated that in national data on traffic levels, certain roadway segments, in particular for local roads, lack accurate traffic counts.[36] State and local officials in two locations also said it was difficult to communicate with the railroad industry, and to get information on train counts, timing and speed of trains from the railroads. BNSF officials told us that they plan to work with communities on rail operating questions and confirm how many trains are moving through a community—and that they will also make them aware of FRA's Rail Crossing Inventory for train count data.[37]

However, FRA depends on railroads and states to voluntarily submit information to the database, and FRA officials stated that the data are not always current.

According to FRA officials, the timeliness of this data should improve, as at the time of our review, FRA officials stated that FRA was in the process of developing an order to require railroads to update information on rail crossings, although the officials did not know when this order would be finalized.

Truck Volumes Are Also Increasing, Causing Impacts in Some Communities

Truck Traffic Also Fell with Economic Downturn of 2008 but Has Rebounded

Similar to rail trends, according to an ATA study, the amount of freight shipped by truck rose from about 10 billion tons in 2000 to a high of almost 11 billion tons in 2006, dropped slightly to about 10.5 billion tons in 2007, and then dropped sharply with the economic recession in 2008, reaching a low of about 8.5 billion tons in 2009 before rising again. According to ATA, trucks transported over 9 billion tons of freight in 2012, representing about 69 percent of all freight tonnage, and ATA officials told us that truck tonnage in 2013 approached but did not yet reach 2007 levels.[38] According to DOT and ATA officials, truck flows—and related freight congestion—are highest in and around major U.S. cities, although the congestion is caused by the combination of trucks, cars, and other vehicles and cannot in most cases be isolated to truck traffic. [39] ATA's research arm, ATRI, has documented that about 89 percent of truck congestion costs can be attributed to about 12 percent of the Interstate highway system mileage. In a 2013 study, ATRI identified significant freight congestion as occurring in particular road segments in major cities in Illinois, Texas, New Jersey, and Missouri. In a 2014 study, ATRI stated that clusters of severe congestion were identifiable in the corridor from Boston to Washington, D.C., as well as Chicago, Texas, and California. According to ATA officials, the congestion in these areas is consistent with where the highest levels of truck-related congestion have been for the past 5 years or more.

ATA and FHWA stakeholders told us that the commodity trend with the most significant effect on truck flow patterns was the increase in oil production in the Bakken region, but that most of this change had occurred on local roads in North Dakota close to the actual oil production. ATRI found that North Dakota experienced the greatest increase in congestion costs on a percentage basis from 2012 to 2013 (about 40 percent). DOT and North Dakota state officials also told us that freight truck traffic in the area had increased tremendously on local roads that are not always included in national traffic studies. North Dakota officials expect high levels of truck traffic to continue in this area over the next 20 years or more. State transportation officials in Pennsylvania and West Virginia, where there is also significant shale oil or gas development, stated that this development has also resulted in increased truck freight traffic in these states, primarily in rural areas.

Truck Traffic on Community Roads Causes Local Traffic Congestion Impacts

DOT, state, and local officials stated that a primary local impact of freight trucks occurs when trucks must exit freeways and travel on roads that pass through communities. Officials we spoke with and a report we reviewed cited impacts including traffic delays and increased difficulty for cars in navigating the streets, including turning and passing, due to the greater size and reduced maneuverability of trucks. [40] Specifically, this report found that the greater size of trucks as compared to cars adds to their congestion effect.

State and local officials we spoke with described high levels of concern related to these truck congestion impacts. In particular, several state and local officials described concerns about traffic congestion impacts caused by freight on roads connecting freeways and freight centers such as ports. In Miami, officials we met with provided us with a study documenting the costs of such impacts. 41 According to this study, trucks comprise 28 percent of the almost 16,000 vehicles that travel to and from the Port of Miami through the downtown streets every day, and the cost of congestion delays associated with freight trucks in the area was estimated to be about $300 million in 2008. Moreover, the traffic-related impacts of freight trucks on roads around East and Gulf Coast ports may increase when the expansion of the Panama Canal is completed in 2015, as both an increase in the number of containers on the larger ships that will be able to pass through the Canal and a potential increase in the number of ships that arrive and depart from ports in this area has been predicted.[42]

Some state and local officials also described traffic congestion impacts of freight trucks in rural or small town areas with major energy extraction. Such areas may not be directly connected to the national freight network by a freeway, so that trucks used in extracting or transporting energy commodities must travel on local roads that cut through small towns and cities. For example, according to North Dakota Department of Transportation officials, between 2007 and 2013, average daily traffic along one main road in the Bakken region increased from under 2,000 to almost 10,000 vehicles, while the percentage of truck traffic increased from 29 to 45 percent (see fig. 9). As described previously, officials in the Bakken region expect oil production and associated truck impacts to continue over the next 20 to 50 years. State officials in Pennsylvania and West Virginia also described negative impacts of increased freight truck traffic on local roads due to shale oil or gas development, including damage to roads and truck congestion impeding the flow of school buses, among other things.

Source: GAO. | GAO-14-740.

Figure 9. Truck Traffic in Watford City, a town near Williston in the Bakken Region of North Dakota, April 2014.

Selected Communities Primarily Used Public Funds to Address Freight-related Traffic Congestion and Some Have Faced Funding Challenges

Project sponsors in our selected communities primarily relied on public funding—provided by federal, state, and local governments and public port authorities—to address freight-related traffic congestion issues; however, some projects we reviewed included private funding. Private funders, in most cases railroads, contributed about 1 percent of the total cost of these projects of $3.2 billion. Railroad companies' contributions to congestion mitigation projects generally amounted to less than 5 percent of the project's total cost and were based on the project's potential operational benefits to the railroad. Some communities we visited have faced challenges funding projects to mitigate freight-related traffic congestion. Among other things, federal programs that can be used to address freight-related traffic congestion do not always align with local congestion mitigation needs, and state and local transportation funds are often limited and needed to operate and maintain existing infrastructure.

The Majority of Funding for Addressing Freight-Related Traffic Congestion in Selected Communities Is Public

Project sponsors in our selected communities, such as state DOTs and public port authorities, primarily relied on public funding to address freight-related traffic congestion issues. Of our 12 selected communities, 9 had ongoing or recently completed congestion mitigation projects. [43] The types of congestion mitigation projects as well as their purposes varied. For instance, we reviewed on-dock rail systems at several ports, which are designed to both enhance freight movements and mitigate congestion due to trucks on local roads, and grade separation projects, which are primarily focused on relieving vehicle delays and traffic congestion. Taken together, these 24 projects cost about $3.2 billion, of which the private sector has contributed about 1 percent. Table 2 below displays the total project cost and the percentage of private funding for each of the mitigation projects we reviewed.

Sixteen of the 24 projects we reviewed have been funded exclusively by the public sector. Of these, five projects were entirely state-funded, three were entirely locally-funded, and one was completed using only federal funds. [44] About half of all projects (11 of 24) included a mix of federal, state, and local funds. Project sponsors reported using various federal funding sources to help pay for freight-related congestion mitigation projects in their communities, including formula funding—such as funds from the National Highway Performance Program, Surface Transportation Program, and the CMAQ program—as well as discretionary funding, such as federal congressionally-directed funding, Transportation Infrastructure Finance and Innovation Act

(TIFIA) credit assistance, and TIGER grants. For example in 2013, almost half of the 52 projects that received a TIGER award primarily addressed freight transportation needs, and among these projects, nearly one-third were intended to mitigate freight-related traffic congestion.

Table 2. Public and Private Funding for Ongoing and Recently Completed Freight-Related Congestion Mitigation Projects in Selected Site Visit Communities (in Millions) as of August 4, 2014

Project[a]	Project's cost[b]	Private funding (as a percentage)
Seattle, WA		
South Spokane Street Viaduct (includes a grade separation)	$157,070	1.6
Duwamish ITS Project	$7,831	1.3
East Marginal Way grade separation	$53,586	2.7
South Atlantic St. grade separation	$42,867	0.0
Spokane, WA		
North Spokane Corridor, Sections 1 and 2	$617,969	0.0
Havana St. grade separation	$23,321	4.6
Chicago, IL		
Grand Ave. grade separation	$44,847	3.5
Belmont Rd. grade separation	$54,903	2.2
CREATE Viaducts	$4,954	0.0
New York City, NY (Metro Area)		
Cross Harbor Rail Freight Program	$118,000	0.0
Express Rail System (On-dock Rail)	$600,000	0.0
Express Rail Overpass and Flyover construction projects	$80,000	0.0
Paulsboro, NJ		
Port of Paulsboro Connector Road	$23,300	0.0
Port of Paulsboro On-dock Rail	$10,400	0.0
Pascagoula, MS		
Port of Pascagoula intermodal improvements	$44,000	11.4
Highway 611/ Industrial Road Grade Separation	$25,637[c]	0.0
Miami, FL		
Port of Miami Tunnel	$721,837	0.0

Table 2. (Continued)

Project[a]	Project's cost[b]	Private funding (as a percentage)
Port of Miami rail access	$46,000	19.6
25th Street Viaduct	$222,082	0.0
Billings, MT		
Billings North Bypass, Phase I	$89,200	0.0
Williston, ND		
Permanent truck reliever route	$139,000	0.0
Highway 2 and 18th Street grade separation	$12,900	0.0
Traffic signal and intersection improvements	$1,700	0.0
Highway 2 and 11th Street intersection reconstruction	$8,700	0.0
TOTAL	$3,150,104	0.7

Source: GAO presentation of information provided by project sponsors. | GAO-14-740

[a]For the purposes of this report, ongoing projects include those that are partially completed, currently under construction, or have identified funding sources. Recently completed projects include those completed since 2007.

[b]The numbers in this column reflect projects' capital costs as reported by project sponsors.

[c]This number does not include the $6.6 million in right-of-way costs associated with the larger Highway 611 Corridor Project ,of which the grade separation is a part.

None of the projects we reviewed have been funded entirely by the private sector. However, eight of the 24 projects included both public and private funding contributions. Most of the projects that included private funding were grade separations (5 of 8) and received contributions from railroad companies (6 of 8). [45] In addition, private sector contributions to these mitigation projects typically amounted to less than 5 percent of the project's total cost, but some projects, such as the Port of Miami's Rail Access Project, included a larger share of private funds (19.6 percent).[46]

See table 2 above for more information about private contributions to specific projects.

Some federal programs encourage public-private partnerships in freight and other transportation projects by giving priority to projects that include innovative partnerships and funding mechanisms. For instance, among DOT's selection criteria for TIGER awards is "partnership"—or the extent to which projects demonstrate strong collaboration among a broad range of participants. Two of the four mitigation projects in our site visit communities that received TIGER grants also included private-funding contributions. For example, the Port Authority of Jackson County Mississippi received a $14 million TIGER grant in 2013 for an upcoming $44 million project at the Port of Pascagoula. This project includes constructing a new marine terminal facility for renewable energy resources, upgrading the port's rail connections, and relocating the rail line that services the port, steps that will allow for the closure of 16 rail crossings in nearby communities. The private entity that is to operate the new marine terminal facility has agreed to contribute approximately $5 million towards the project. In addition, as part of determining eligibility for its TIFIA Credit Assistance Program, DOT pursuant to statutory requirements assesses the extent to which TIFIA assistance would foster innovative public-private partnerships, attract private debt or equity investment, and reduce the contribution of federal grant assistance to the project.[47] One project we reviewed—the Port of Miami Tunnel—was financed in part by a $341 million TIFIA loan. DOT awarded the loan to a private entity to help finance the construction of the tunnel, which provides direct access to the port from the nearby interstates and helps reduce port-related traffic congestion on downtown streets. Under this arrangement, the private partner financed, designed, and constructed the tunnel and will operate and maintain it over the next 30 years. The public sector (Florida DOT, Miami-Dade County, and the City of Miami) agreed to repay all of the capital, operations, and maintenance expenses associated with the tunnel by making periodic, performance-based payments to the private partner.

Railroads' Contributions for Mitigating Freight-Related Traffic Congestion Are Primarily Driven by the Potential Operational Benefits

As noted above, private contributions to projects we reviewed aimed at mitigating freight-related traffic congestion typically came from railroad companies.[48] Railroad representatives we spoke with told us that their companies rely on cost-benefit analyses to make decisions about whether and how much to contribute toward public-private partnerships, and that the contributions generally align with a project's potential operational benefits to the company. For example, CSX officials told us that, among other factors, they consider the potential operational benefits to the company when determining whether to participate in a public-private partnership. They noted that if they find that a project has value to the company and the overall partnership makes sense, CSX will provide funding commensurate with its private benefit. In their view, contributions beyond that amount would be paying for benefits to the environment or the community, but not the railroad. Similarly, as part of its framework for public-private partnerships, Union Pacific Railroad includes the following principles: 1) the railroad pays for private benefits that will accrue to the company and 2) public entities pay for the benefits that are enjoyed by the public at large through public or general revenue sources.

Federal regulations pertaining to railroads' contributions for grade crossing elimination projects—such as grade separations or crossing closures—also reflect the estimated operational benefit of these projects to the railroads. Specifically, FHWA's regulations require railroad companies to contribute 5 percent toward grade-crossing elimination projects that receive funding from FHWA's Railway-Highway Crossings Program. 49 According to DOT officials, the 5 percent railroad contribution to grade-crossing elimination projects is based on a cost-benefit analysis and represents the estimated benefit to the railroad of no longer having to operate and maintain the crossing.[50]

Railroad companies we spoke with told us that FHWA's 5 percent requirement generally reflects an appropriate contribution toward grade separation projects for railroads, since these projects typically do not provide significant operational benefits to their companies. For instance, according to Chicago Transportation Coordinating Office representatives, although grade separation projects result in some benefit to the railroads, the bulk of the benefits accrue to the community in the form of safety improvements and congestion relief. In addition, BNSF representatives told us that the company typically expects to contribute the required 5 percent to grade separation projects funded by the Railway-Highways Crossings Program, since crossing closures generally do not produce significant financial benefits to the company. However, according to BNSF officials, if the results of the company's cost-benefit analysis indicate that a crossing closure would produce benefits greater than 5 percent, either through cost avoidance or increased revenues, BNSF would be willing to negotiate a contribution greater than the minimum amount required.

In certain circumstances, STB—which issues licenses to railroads for rail construction and acquisition projects—can order a railroad to contribute more than 5 percent towards a grade separation project as a condition of its license approval. STB can impose this license condition on a railroad only if:

- the railroad is applying for a license from STB to construct a new line that is extending into new territory, or is merging, consolidating, or acquiring control of a rail carrier that is under the jurisdiction of STB; and
- the transaction is consistent with the public interest.[51] STB makes the determination based on whether its environmental review of the proposed construction or acquisition would result in environmental impacts that justify a grade separation.[52]

For example, STB used this authority to order Canadian National Railroad to pay for most of the costs of two grade separation projects (67 percent and 78.5 percent respectively) as part of its acquisition of the Elgin, Joliet, and Eastern Railway rail line.[53] According to STB officials, in this case, their review determined that the railroad would receive substantial economic benefits from STB's approval of the acquisition and the community would experience significant environmental impacts due to the additional trains operating on the line. As a result, STB determined it was appropriate to require the railroad to incur the majority of the grade separations' cost. STB officials informed us that during the environmental review process, rail applicants often offer voluntary environmental mitigation measures to respond to community concerns, including freight-related traffic congestion. In several cases, STB has not had to impose grade separation conditions due to railroads' voluntary mitigation. STB officials emphasized that STB has no authority to require railroads to pay for grade separations along existing lines unless they are part of a construction or acquisition project that necessitates an environmental review by STB. In addition, they told us that STB cannot mitigate freight-related traffic congestion issues in communities by regulating changes in rail traffic levels because freight railroads generally have the discretion and authority to determine how to route their trains.[54]

Communities Face Several Challenges Funding Projects to Address Freight-Related Traffic Congestion

Although the public sector contributed the vast majority of funding for projects aimed at mitigating freight-related traffic congestion in our selected communities, securing funding for these projects can be challenging for state and local transportation officials. For instance, local officials in Glendive, Montana, and neighboring Baker, Montana, told us that their communities have experienced significant increases in truck and rail traffic due to the recent development of crude oil development in the Bakken region, but the officials have not identified funding for mitigation projects. Several of the communities we visited with ongoing or recently completed projects also had additional planned projects that were not yet funded. For example, Chicago's CREATE Program lacks about $2.0 billion to complete 22 remaining projects, 13 of which are grade separations.

At the federal level, funding programs that can be used to address freight-related traffic congestion have broader goals and objectives that do not always align with communities' congestion mitigation needs (see table 1 above for descriptions of these programs). For instance, DOT's Railway-Highway Crossings Program—which provides funding to states for highway-rail grade crossing improvements—is focused on addressing safety issues, not congestion. According to DOT officials, Railway-Highway Crossings Program funds can be used to address freight-related traffic congestion so long as the congestion also presents a safety issue.[55] However, stakeholders noted that communities' congestion and safety issues do not always intersect. For instance, Washington State DOT (WSDOT) officials told us that while the Railway-Highway Crossings Program's focus on safety is important, some communities face major congestion issues that cannot be easily incorporated into a safety-related project. They also noted that the concerns about highway-rail grade crossings that WSDOT hears from communities across the state primarily involve congestion issues, not safety issues. In cases when there is overlap between congestion and safety issues, it can be difficult for states to fund qualifying grade separation projects using these funds. The average cost of the seven grade-separation projects we reviewed was $37 million, yet most states receive less than $10 million from this program annually and 50 percent of states' funds must be spent on grade-crossing warning devices. [56]

At the state and local level, several transportation agencies told us that much of their funding is needed to operate and maintain existing infrastructure.

As a result, very little funding is available to complete new projects or to serve as a local match for federal-funding programs.[57] For instance, New Jersey DOT (NJ DOT) officials told us that maintaining the state's existing infrastructure— much of which is old and in need of repair—is the state's priority and takes up most of NJ DOT's funding.

Consequently, the number of new projects NJ DOT can undertake to address congestion is low. Similarly, an official in Billings, Montana, told us that local transportation agencies are reluctant to undertake major new projects, including those that are aimed at alleviating freight-related traffic congestion, because funding is limited and one major project can tie up 5 or 6 years worth of available funds.

In addition, local transportation officials in Miami stated that it is almost impossible for many communities to come up with a local match for freight-related projects because most local revenues are used to operate and maintain existing infrastructure and fund other priorities.

We have previously found that freight projects may not compete well with other types of transportation projects for limited available public funds because their benefits are not always obvious to the public. 58 We have also found that public planners are wary of providing public support for projects that directly benefit the private sector, a wariness that may make it less likely that metropolitan planning organizations or state departments of transportation will prioritize freight projects.[59]

Although projects to mitigate freight-related traffic congestion in communities may have more discernible public benefits than other freight projects, other factors can affect their ability to compete or receive priority for funding. For example, local transportation officials in Spokane, Washington, told us that progress on the Bridging the Valley project has stalled because they have been unable to secure funding for the project and operational commitments from the two railroad companies involved. Local officials in Billings stated that efforts to resolve long-standing traffic congestion issues at highway-rail grade crossings in the city have been stalled both by a lack of funding and by a lack of consensus on the best alternative for addressing these issues. In addition, officials with the Port Authority of New York and New Jersey told us that it is difficult for the Port Authority to invest in roads that connect highways to ports because—unlike the tunnels and bridges it operates—it doesn't own or receive revenue from these roads. To address challenges associated with funding connector roads, the National Freight Advisory Committee recommended in June 2014 that DOT develop federal-funding programs that support and prioritize connectors that are part of regionally or nationally significant freight networks.

Data limitations may also make it challenging for project sponsors to fully document the impacts of freight movements and compete for limited available funding.

As described previously, communities we visited had varying levels of data on local impacts of freight-related traffic congestion, with some communities lacking specific information on train or vehicular movements to effectively measure these impacts.

Moreover, national data in this area also has limitations—specifically related to the timeliness and completeness of the data at the local level. These limitations are described in more detail later in this report.

DOT's Efforts to Implement MAP-21 Freight Provisions Incorporate Some Consideration of Community Impacts but Do Not Establish a Clear Federal Role

Although communities have faced funding challenges in mitigating freight-related traffic congestion, officials from freight trucking, rail, intermodal, and shipping associations or companies and from various state or local transportation agencies we contacted and the American Association of State Transportation Officials agreed that MAP-21 was a positive step in defining a national focus on freight. As previously described, MAP-21 does not establish as an explicit goal that the national freight policy address community impacts of freight, such as local traffic congestion impacts. However, MAP-21 does include a goal to reduce congestion— which could incorporate local traffic congestion caused by freight flows— and a goal to reduce the environmental impacts of freight movement on the national freight network, which relates to a type of community impact that can be caused by freight congestion.

MAP-21 also includes as a required element of a national freight strategic plan the identification of best practices to mitigate the impacts of freight movement on communities. Within this framework of limited direction related to community impact issues, DOT has taken some steps to address local impact issues in its efforts to implement MAP-21's freight provisions, but to date has not fully integrated local impact issues, including freight-related traffic congestion, or established a clear federal role in this area, in part due to data limitations.[60]

With Mileage Limitations and No Clear Purpose Established, DOT's Draft Highway Primary Freight Network Omits Some Road Types with Local Freight Impacts

As a component of establishing a national freight network, MAP-21 requires DOT to establish a highway primary freight network consisting of not more than 27,000 centerline miles of existing roadways that are most critical to the movement of freight.[61] The significance of the 27,000 mileage limitation is not clear. In November 2013, DOT issued for comment its draft primary freight network. However, DOT and other national and state freight stakeholders that commented on the draft network raised concerns that the lack of a defined purpose for the network in MAP-21 and the mileage limit requirement—that it be limited to 27,000 centerline miles—had resulted in a network that omits the types of roads where local traffic congestion impacts of national freight movements are often experienced. MAP-21 established a general vision for the national freight network,[62] stating that the national freight network is to assist states in strategically directing resources toward improved system performance for efficient movement of freight on highways, including the National Highway System, freight intermodal connectors, and aerotropolis transportation systems.[63] MAP-21 also includes eight factors that DOT is to consider in establishing the highway primary freight network.[64] While these factors do not specifically mention local traffic congestion impacts of freight movement, they do incorporate several areas where such impacts are typically experienced, such as land and maritime ports of entry, access to energy exploration or production areas, and population centers.

However, DOT stated in its comments that its ability to designate a highway primary freight network that successfully takes into account all eight factors was hampered by the lack of a clear purpose established for the primary freight network and by the mileage limit requirement of 27,000 centerline miles, in which the mileage of each road included in the network is counted against this mileage limit. In contrast to this centerline approach, DOT has at times used a corridor approach in past transportation work, in which the mileage of a transportation corridor (such as the corridor between Washington, D.C., and Baltimore, Maryland) would be counted towards the mileage of a network, while allowing multiple roads—or even rail lines, if the corridor was multimodal—along that corridor to also be included as part of a network.

In fact, when it issued the draft primary freight network of 27,000 centerline miles for comment, DOT also identified a broader draft network of more than 41,000 centerline miles that, according to DOT's analysis, would be necessary to establish a primary freight network that consisted of a comprehensive, connected set of roads for transporting goods efficiently on highways throughout the nation. The comment period on the draft primary freight network closed on February 15, 2014. According to DOT officials, DOT is in the process of reviewing comments on the highway primary freight network.

Several organizations representing public or private freight-related officials also commented that the prescribed mileage limit and centerline approach had contributed to the selection of a network in which the roads connecting ports or other freight centers to the Interstate Highway System were not included. For example, according to the National Freight Advisory Committee, few ports are connected to the proposed network, including some of the busiest ports. One analysis found that 82 percent of the over 150 policy-related comments submitted stated that freight intermodal connectors—such as roads connecting ports to freeways— should be included in the highway primary freight network.[65] The National Freight Advisory Committee commented that the draft primary freight network fell short in part because it generally does not include these connector roads, and the roads that trucks use to carry freight from and into communities—the roads where community impacts are most directly experienced. Similarly, the I-95 Corridor Coalition criticized the draft, stating that it omitted roads connecting major metropolitan areas and roads serving important resource areas, such as those connecting the New York State and Pennsylvania gas and oil fields to the region's processing and population centers. These types of roads, which were largely not included in the network in order to meet the mileage limit, are all areas in which freight movement is likely to cause local traffic congestion. In addition, the American Association of State Highway and Transportation Officials stated that after seeing DOT's draft highway primary freight network, it became apparent that a 27,000 mile network based on centerline miles is far too small—and it urged Congress to adopt a corridor-based approach which incorporates multiple highway facilities rather than highway centerline miles.

In addition, the highway-focused and centerline approach precludes including those rail line corridors where high levels of rail traffic going through communities are likely to lead to local traffic impacts at highway-rail grade crossings. Railroad and DOT officials we spoke to told us that grade crossing separations are considered largely a highway, not a railroad issue. If those roadways where highway-rail grade crossings cause delays are not included in the highway primary freight network, it may limit states' flexibility to mitigate them if, for example, funding were to be attached to the primary freight network.

DOT and the National Freight Advisory Committee expressed the concern that without clear goals for the highway primary freight network—and in the context of the mileage limitation—it was difficult to select or comment on which roads should be included in the network. As stated above, Congress provided a general overarching vision for the national freight network—to assist states in strategically directing resources toward improved system performance for efficient movement of freight on highways. However, as the National Freight Advisory Committee stated in its comments on the draft primary freight network, the intended purpose of the primary freight network was not defined, including such questions as how the primary freight network will be used and whether it will be used to prioritize needs, justify future investment, or to develop rules or regulations specific to the freight industry. DOT stated that without a better understanding of the goals for the highway primary freight network, it was challenging to gauge whether the resulting network would meet future public planning and investment needs.

Leading practices in capital planning emphasize that vision is a critical success factor—and that in establishing a vision, a defined purpose and goals are critical for determining which areas should receive increased emphasis and funding and which areas should remain stable or receive reduced emphasis.[66] In addition, our prior work has identified a framework of principles that can help inform Congress in assessing federal surface transportation programs. These principles include creating well-defined goals based on identified areas of national interest and establishing and clearly defining the federal role in achieving each goal in relation to the roles of state and local governments and other stakeholders.[67] In line with this framework, in prior work, we found that clearly defining the federal interest allows policymakers to clarify the goals for federal involvement and can clearly define the roles of federal, state, and local government in working toward each goal.[68] The lack of a clearly defined purpose for the highway primary freight network, combined with the mileage limit imposed, resulted in a draft network that omits certain types of freight-related roads, such as roads connecting ports to freeways, while including others, such as major Interstate freeways with heavy truck volumes, without a clear understanding that this was the network Congress intended or of the ramifications of these choices. According to DOT officials, the omission of certain types of freight-related roads from the national freight network—of which the primary freight network is one part—was driven by several factors, including the requirement that all interstate highways be included in the national freight network. DOT stated that the non-inclusion of roads connecting ports to freeways was due to the fact that MAP-21 allows rural freight connectors to be designated by the states as part of the national freight network but did not allow urban freight connectors to be designated by the states as part of the national freight network. Moreover, DOT cited a lack of data on these roads connecting ports to the Interstate Highway System that makes it difficult to identify at the federal level which ones should be a part of the national freight network.

DOT's surface transportation reauthorization proposal, GROW America, which was introduced in the House of Representatives in June 2014, could address concerns about establishing a clear purpose for the primary freight network and the mileage limit requirement of 27,000 centerline miles. GROW America would replace MAP-21's highway primary freight network with a multimodal national freight network—with no mileage limit—that has a defined purpose to inform public and private planning, prioritize for federal investment, aid the public and private sector in strategically directing resources, and support federal decisionmaking to achieve national freight policy goals.[69] DOT stated that this proposal would also eliminate the causes described above that led to a lack of inclusion in the primary freight network of roads connecting ports to highways. Whether Congress adopts DOT's legislative proposal or otherwise, without establishing a clear purpose for the national freight network and revising the mileage limit requirement of the primary freight network so that more freight-significant road segments could be included in it—including those where community impacts of freight are often experienced—the federal role in freight-related traffic congestion will remain unclear.

With Limited Guidance from MAP-21, DOT Has Not Yet Defined the Federal Role in Addressing Local Freight Impacts and Lack of Data Hampers Effort

In the context of limited direction in MAP-21 related to community impact issues, to date, DOT's efforts to implement certain freight requirements in MAP 21 do not clearly establish a federal role related to local impacts such as freight-related traffic congestion by identifying goals, objectives, or measures in this area. As previously described, MAP-21 does not establish as an explicit goal that the national freight policy address community impacts of freight, such as local traffic congestion impacts, although it does include a goal to reduce congestion—which could incorporate local traffic congestion caused by freight flows—and a goal to reduce the environmental impacts of freight movement on the national freight network, which relates to a type of community impact that can be caused by freight congestion. MAP-21 does include as a required element of the freight strategic plan the identification of best practices to mitigate the impacts of freight movement on communities. [70]

In spite of this limited direction from MAP-21, DOT has taken some steps to address local impact issues in implementing MAP-21's freight provisions. In 2012, DOT issued interim guidance to states on developing state freight plans.[71] MAP-21 states that the state freight plans are to provide a comprehensive plan for the immediate and long-range planning activities and investments of the state with respect to freight. Furthermore, DOT outlines in its interim guidance to states that it plans to rely significantly on state freight plans in preparing the *National Freight Strategic Plan*. Several state, local, and private-sector freight-transportation officials we spoke with said state freight plans were a good step to encourage states to specifically plan for freight within the state and to facilitate their work with corridor-focused groups.[72] Moreover, the National Freight Advisory Committee recommended that DOT should require and fund the development of state freight plans and set up mechanisms to ensure that state departments of transportation interact with all transportation modes, users, regional and multi-state agencies, and metropolitan planning organizations.[73]

State freight plans may provide an opportunity to better define roles and prioritize freight- planning efforts. As described above, our prior work has identified a framework of principles that can help inform Congress in assessing federal surface transportation programs. These principles include creating well-defined goals based on identified areas of national interest through an examination of the relative priority of existing programs and establishing and clearly defining the federal role in achieving each goal in relation to the roles of state governments and other entities.[74] DOT's interim guidance on state freight plans lists "reducing adverse environmental and community impacts of the freight transportation system" as one of six national goals and generally discusses in the guidance that states should develop measures of conditions and performance for all six goals. According to DOT officials, while reducing the adverse community impacts of freight transportation was not included as one of the goals of the national freight policy in MAP-21, DOT included reducing adverse community impacts as one of the goals in its interim state freight plan guidance in part because of the requirement that the *National Freight Strategic Plan* discuss best practices to mitigate the impacts of freight movement on communities and in part because of DOT's commitment to enhancing livability and quality of life. DOT stated that by including such impacts as a goal, it went further than required by MAP-21. DOT also stated that as required by MAP-21, it plans to incorporate a discussion of best practices to mitigate the impacts of freight movement on communities in the *National Freight Strategic Plan,* which is due in 2015.

DOT's inclusion of community impacts as one of six goals in the interim guidance on state freight plans is a positive step that could help further the understanding of the federal role in considering such impacts as well as encourage states to incorporate information in state freight plans that could be helpful to DOT's efforts to include best practices in this area in the *National Freight Strategic Plan*, as required by MAP-21. However, DOT's interim guidance to states on state freight plans did not identify specific goals, objectives, or performance measures for addressing freight-related congestion in local communities, and does not provide details on how states could consider documenting or addressing community impacts such as freight-related traffic congestion.[75]

According to DOT officials, DOT did not include details in its state freight plan guidance on how states could consider documenting or addressing community impacts because MAP-21 does not require state freight plans to consider community impacts, and because states are not required to adopt federal goals and are not required to prepare a state freight plan. Moreover, DOT officials stated that it should be up to the states to identify goals, objectives, and performance measures in their state freight plans. However, without specific guidance related to how to consider community impacts amongst other freight priorities, it may be difficult for those states with concerns related to the local traffic congestion impacts of freight to fully incorporate these concerns and issues in their state freight plans. This difficulty, in turn, could reduce the usefulness of state freight plans to DOT in considering these issues for the *National Freight Strategic Plan* in order to more clearly articulate the federal and state roles in this area.

In addition, DOT has begun work on the required *Freight Transportation Conditions and Performance Report*, which is due in October 2014, and which will also eventually feed into the *National Freight Strategic Plan.* According to DOT officials, for the *Freight Transportation Conditions and Performance Report*, DOT is developing performance measures in six areas, one of which—reducing adverse environmental and community impacts—relates to the impacts of freight on local communities.[76] However, DOT officials stated that they have not established a clear vision of the federal role in this area because of a lack of national data on these issues.[77] For example, DOT officials stated that while they have been working to develop quantitative measures in each of the six areas established for the *Freight Transportation Conditions and Performance Report*, they do not have national data on adverse environmental and community impacts and will probably discuss these issues anecdotally.

As a result, in the *Freight Transportation Conditions and Performance Report*, DOT is not planning to identify specific goals, objectives, or measures for addressing freight-related congestion in local communities.

DOT also expressed concerns about the availability of national freight data in its comments on the draft primary freight network. Specifically, DOT expressed concern that data limitations could reduce the likelihood that certain types of roads where freight-related congestion impacts are likely to be felt would be included in the highway primary freight network. DOT stated that the data utilized for the development of the draft primary freight network comprises the best information available on freight behavior at a national level.[78] However, DOT commented that urban and rural areas that are the site of significant freight facilities where highway freight intersects with other modes, such as ports, do not always show up well in the available national data sets it used for developing the highway primary freight network.

In June 2014, the National Freight Advisory Committee cited the lack of data for monitoring and analyzing the freight system as a barrier to improved system performance and made several recommendations related to improving data. These recommendations included, among other things, that data collection efforts should be tailored to performance measures that are in line with specific outcomes that DOT and Congress want to obtain, with increased emphasis on the multimodal national freight system, and that DOT should partner with objective third party organizations to facilitate data collection agreements with private industry.[79]

Leading practices in capital planning emphasize that another important success factor in effective decisionmaking is the availability of good information. Good data and information systems are essential to supporting sound capital planning and decision-making. The data and information provided by well-planned information systems give organizations the ability to build comprehensive measures, collect relevant data, and perform analyses that can be used to support strategic as well as operational budgeting decisions.[80] As previously described, we also found that data limitations reduced our ability to analyze highway freight trends—information that could be key to helping states and the federal government establish relevant goals and prioritize mitigation efforts in these areas. We found the following causes of these data limitations:

- While states are required to annually submit information on highway traffic counts for DOT's Highway Performance Monitoring system (a national level highway information system that includes data on the extent, condition, performance, use, and operating characteristics of the nation's highways), some DOT officials cautioned against using this information as a measure of freight flows because of concerns that information on truck traffic percentages is not consistently recorded. Moreover, the truck volume information submitted by states is not separated out by freight truck versus other types of trucks, but rather by vehicle class, such as light truck, single unit, or combination unit. DOT does present data on national freight truck flows through its freight analysis framework, which, according to DOT, integrates data from a variety of sources to create a comprehensive picture of freight movement among states and major metropolitan areas by all modes of transportation. DOT considers this data to be reliable, but this framework depends, in part, on Census data and additional analysis by DOT, and is updated only every 5 years, with over a 2-year reporting lag. In other words, the currently available freight analysis framework data is from 2007, and the updated freight analysis framework data, which will be based on 2012 data, is not expected to be released until 2015.
- While DOT officials stated that they consider the Highway Performance Monitoring System data on vehicle counts for general vehicular (not specifically truck) traffic to be reliable for roads that receive federal aid, very limited data on roads off the national highway system is collected; thus, DOT officials do not consider Highway Performance Monitoring System data a good tool with which to understand specific local impacts of freight traffic.
- According to some DOT, state, and local officials, private railroads are often unwilling to provide detailed information on train counts, lengths, speed, and times to local communities to help them quantify traffic impacts at highway-grade rail crossings. Some state officials had received much of this specific information for certain highway-rail grade crossings, while others we spoke to said they had been unable to get this information. As previously described, BNSF officials stated they plan to work with communities to provide

relevant information, and that they refer communities to FRA's Grade Crossing Inventory, but this inventory includes only average daily train numbers and the timetable (not actual) speed, and, as previously described, may not be current.

DOT officials stated that in the absence of reliable national data related to potential adverse environmental and community impacts of freight, DOT will likely discuss this anecdotally in the *Freight Transportation Conditions and Performance Report* being developed. They stated that they planned to outline categories of community impacts that might be significant, including noise, heavy freight traffic on community streets, and railroads blocking highway-rail grade crossings and interfering with operations of emergency vehicles. DOT's efforts may represent the beginning of defining a national role related to the traffic-related congestion impacts of freight on communities. For example, the categories DOT told us they were considering align well with some of the areas that stakeholders told us were most problematic in terms of impacts on the community and with the areas where communities' mitigation efforts have been focused. According to DOT officials, DOT plans to attempt to define national data limitations in the *Freight Transportation Conditions and Performance Report* and to describe what actions DOT is taking to reduce these limitations. However, DOT stated that it was not planning to put forward a strategy to improve the availability of national data to quantify and assess freight trends and impacts on local traffic congestion because of the difficulty of doing so. DOT stated that national data are, by definition, national, and cannot quantify impacts on local traffic congestion. DOT also stated that the question of how to aggregate local data to produce national measures is a difficult one, and it is not yet clear how this can be accomplished. We understand that improving the availability of national data to quantify and assess freight trends and impacts on local congestion is a difficult and complex task. Moreover, we understand that such an effort could involve some combination of improving the level of detail of national data or aggregating local data to produce national measures, and that efforts in each of these areas are likely to involve resolving many underlying issues. At the same time, DOT has developed guidance to help improve the granularity of national data in ways that could help improve the national understanding of local traffic congestion impacts. For example, in its current guidance to states on the Highway Performance Monitoring System (under frequently asked questions), DOT includes information on how states can work to improve the information they provide for this national data set on the percentage of trucks on road sample sections. In our Executive Guide on the Government Performance and Results Act (GPRA), we found that a key step for agencies to take to become more results-oriented was to define clear missions and desired outcomes—corresponding to the requirement in GPRA for federal agencies to develop strategic plans.[81] Without a strategy defining clear missions and desired outcomes related to improving data that would help quantify freight-related traffic congestion, DOT may miss the opportunity to build on its current data-improvement efforts and identify future data-improvement efforts that could help establish and clarify the national role in this area.

In addition, as previously described in this report, we found data limitations at the state or local level that reduced state or local officials' ability to quantify rail-related freight impacts on local communities. As a result, we found that communities we visited had varying levels of quantified information on impacts such as traffic delay times or costs. The inconsistency and incompleteness of such information at the state or local level is likely to reduce the ability of states to appropriately prioritize efforts to mitigate such impacts against other types of funding priorities and measure progress towards goals—and could also hamper DOT's efforts to consider these issues from a national perspective. Without reliable national and state data on the community impacts of freight, measuring the extent of freight-related traffic congestion and tracking the progress of efforts to mitigate the impacts will be difficult to do. However, DOT's interim guidance on state freight plans did not provide specific suggestions as to what data states could collect, report, or analyze in order to improve their ability to prioritize impacts of freight on local traffic congestion or other freight-related issues.

DOT's efforts as previously described do not yet fully establish the federal role in this area, which could limit the usefulness of the eventual *National Freight Strategic Plan* in laying out a strategy related to these issues. In our *Executive Guide on the Government Performance and Results Act* (GPRA), we found that a key step for agencies to take to become more results-oriented was to define clear missions and desired outcomes— corresponding to the requirement in GPRA for federal agencies to develop strategic plans.[82] The *National Freight Strategic Plan* is therefore an appropriate document for DOT to incorporate the key principles previously described for assessing surface transportation programs, including creating well-defined goals based on identified areas of national interest, which involves examining the relevance and relative priority of existing programs in light of 21st century challenges and identifying emerging areas of national importance.[83] The extent to which local traffic-congestion impacts of freight are of national importance has not been clearly established. However, it is clear that these impacts are caused by national freight movement that has national benefits—but, especially in the case of pass-through traffic—does not necessarily have benefits for the communities experiencing the congestion or other negative impacts. Moreover, as described earlier, it appears that freight traffic may increase in coming years, particularly in certain states and along certain corridors, and these increases, if they occur, are likely to increase certain national benefits while also increasing negative local impacts in affected communities. The development of the *National Freight Strategic Plan* provides an opportunity to more clearly focus the federal role in this area. As we have concluded in prior work, without a clearly defined federal role, policymakers lack clarity on how to allocate scarce federal resources according to the level of national interest.[84]

CONCLUSION

Freight-related traffic congestion affects communities of all sizes in the U.S. In some locations, communities have struggled for decades to address long-standing freight-related traffic congestion issues due to funding and other challenges; elsewhere, such as in North Dakota's Bakken Region, these impacts have emerged relatively recently. Given that freight volumes are expected to continue to increase over the next 25 years and beyond, a growing number of communities will likely be seeking to devise and fund solutions to mitigate the impacts of freight-related traffic congestion.

The implementation of MAP-21's requirements provides an opportunity to consider the extent to which there should be a national role in addressing freight-related traffic congestion. Although it is known that national freight movements contribute to local freight-related traffic congestion, neither MAP-21 nor DOT has yet established a clear federal role in mitigating or helping communities address these impacts. Defining this role is likely to be more important in coming years if freight traffic, in particular related to the movement of oil and coal for export by train and the intermodal movement of goods in and out of ports by truck or train, increases as expected. However, DOT's efforts related to the development the *National Freight Strategic Plan*—including its interim guidance to states on state freight plans—so far have not established a clear federal role by identifying specific goals, objectives, and measures for addressing freight-related congestion in local communities. These limitations could limit the usefulness of the eventual *National Freight Strategic Plan* in laying out a strategy related to these issues.

Moreover, underlying any efforts to address freight-related traffic congestion are a number of data limitations both national and local, which if resolved, may assist in better defining the extent of freight-related congestion and its impacts and also assist in prioritizing projects to mitigate such impacts at both the state and national level. Finally, in its draft primary freight network, DOT was hampered in attempting to take into consideration all eight factors Congress provided as guidance by the lack of a clear purpose established for the highway primary freight network and by the mileage limit imposed. Consequently, it is unclear how the primary freight network will eventually be used or what Congress intended to be included or excluded from the primary freight network. Clarifying the purpose and goals of the network and determining the extent to which local roads and connectors should be included—and the extent to which the mileage requirements prohibit the inclusion of freight-important roads, including those where communities often experience freight related traffic congestion—would help to establish the federal role in freight-related traffic congestion.

Matter for Congressional Consideration

In reauthorizing the federal highway program, Congress should consider establishing a clear purpose for the national freight network and primary freight network that incorporates inclusion of the types of roads where communities are likely to experience significant freight-related traffic congestion, and, as relevant to this purpose, consider revising certain requirements such as the mileage limit of 27,000 miles or changing the requirement from a centerline to a corridor approach.

Recommendations for Executive Action

In order to clarify the federal role related to freight-related local traffic congestion, we recommend that the Secretary of Transportation take the following two actions in implementing MAP-21 and any subsequent reauthorization:

- in its final guidance on state freight plans, incorporate additional information to help states define and prioritize local community impacts of national freight movements, including traffic-congestion impacts, and to establish what data could be consistently collected and analyzed in order to prioritize impacts of freight on local traffic congestion;

- include in the *National Freight Strategic Plan* a written statement articulating the federal role in freight-related local congestion impacts, by clearly identifying potential objectives and goals (under the general area DOT has established for the *Freight Transportation Conditions and Performance Report* of reducing adverse environmental and community impacts) for mitigating local congestion caused by national freight movements and the type of role federal and state stakeholders could play in achieving each objective and goal, and including a written strategy for improving the availability of national data needed to quantify, assess, and establish measures on freight trends and impacts on local traffic congestion.

AGENCY COMMENTS AND OUR EVALUATION

We provided a draft of this report for review and comment to DOT. We received written comments from DOT's Assistant Secretary for Administration.

In its letter, DOT stated that it is making significant progress in addressing community congestion impacts and the development of a national freight strategy. DOT stated that its proposed GROW AMERICA Act continues to build on the freight provisions in MAP-21 by identifying and mitigating congestion and other community impacts resulting from the transportation of freight. DOT also stated that in developing the *National Freight Strategic Plan*, DOT has heard from the National Freight Advisory Committee and other stakeholders recommending that DOT articulate a clear federal role in the movement of freight, including addressing first and last mile connectors in both urban and rural areas where community impacts are often most acutely felt. DOT also stated that it is incorporating these and other comments in the *National Freight Strategic Plan*, undertaking research, and developing data and tools to support an outcome-oriented, performance-based approach to freight projects and planning.

In a draft version of this report provided to DOT, we recommended that DOT issue a legislative proposal that would address concerns raised in our report about the highway primary freight network, including the lack of a clear purpose for the primary freight network and the primary freight network's mileage limit requirement of 27,000 centerline miles. In technical comments DOT provided to us after reviewing the draft, DOT stated that its surface transportation reauthorization proposal, GROW America, which was introduced in the House of Representatives in June 2014, would address these concerns through the establishment of a multimodal national freight network that would replace the primary freight network—and that had a defined purpose and no mileage requirement. We reviewed the GROW America proposal and agreed that the multimodal national freight network proposed to replace the primary freight network would address the concerns raised in our report about the primary freight network. We therefore revised our draft accordingly, including removing the recommendation and adding a Matter for Congressional Consideration related to the national freight network and the primary freight network.

Sincerely yours,
Susan A. Fleming
Director,
Physical Infrastructure Issues

APPENDIX I: OBJECTIVES, SCOPE, AND METHODOLOGY

This report addresses the following objectives: (1) how U.S. rail and truck freight flows are changing, and the extent to which freight-related traffic congestion is reported to impact local communities; (2) how communities have funded efforts to mitigate freight-related traffic congestion, and what funding challenges, if any, communities report facing; and (3) the extent to which DOT's efforts to implement freight-related provisions of MAP-21 have addressed local freight-related traffic congestion.

To determine how U.S. rail flows are changing, we analyzed Surface Transportation Board (STB) data on rail carloads from 2007 to 2012, the most recent year available, including information on all carloads and on intermodal carloads, and on carloads of selected commodities, including coal and crude oil. We selected carloads as a measure because of its relevance to traffic congestion at highway-rail grade crossings. We chose to analyze STB data beginning in 2007 after being advised by STB officials that to examine recent trends, 2007 was more indicative than 2008 of rail traffic before the recent U.S. economic recession. We focused on trends related to intermodal, coal, and crude oil transportation by rail because they were emphasized by STB and AAR officials as having a significant impact on changes in rail traffic in recent years. In addition, intermodal and coal carloads are the only two product types to make up at least 20 percent of all rail carloads in 2012, and, while crude oil represented only about 1 percent of all rail carloads in 2012, the percentage increase in crude oil carloads from 2007 to 2012—of approximately 3,188 percent—was far higher than the percentage change of any other commodity. We also looked at trends in carloads at the national and state level, using the STB data's state-through indicator as the measure of whether a carload was in a state. We reviewed the reliability of STB's data by reviewing related documentation, examining the data for irregularities, and interviewing agency officials about their data collection procedures and controls for maintaining the data. Carload figures are considered estimates because they are based on data sampling rather than complete reporting. We estimated that for 2007- 2012, the margin of error for carloads of all commodities we examined was less than 5 percent except crude oil, which had a higher margin of error until 2012. Crude oil carloads, which existed in small numbers in 2007-2011, had for those years a declining margin of error that started at 26 percent in 2007 and went down to about 8 percent in 2011. For 2012, the margin of error for crude oil carloads was less than 5 percent. We determined that these data were sufficiently reliable for the purposes of our review. We also received information from the Association of American Railroads (AAR) on rail carload levels in 2013, according to AAR data.

To provide context on freight flow volumes in the U.S., we reviewed the Department of Transportation's (DOT) methodology for a map it had published showing freight flow tonnage on highways, rail, and inland waterways for 2010. This map used a combination of DOT's 2007 Freight Analysis Framework data, updated through additional analysis involving DOT's Highway Performance Monitoring System (HPMS) truck flow data and economic information in order to present information on highway freight flows; STB data on rail volumes in 2010 as flowed over the rail network by DOT's Federal Railroad Administration (FRA); and Army Corps of Engineers' 2010 data on freight tonnage on inland waterways. We investigated whether we could update this map to 2012 but learned that we could not because 2010 was the most recent year available for sufficiently reliable information on highway freight flows. We therefore analyzed these data to present a map of 2010 freight flows. We reviewed the reliability of these data by examining documentation from the agency, interviewing knowledgeable agency officials about the data collection procedures, and discussing with FRA officials their procedure for flowing STB rail data on rail lines. We concluded that these data were sufficiently reliable for the purposes used in our report.

To determine how truck freight flows are changing, we investigated whether we could use a federal data source on truck flows, but on reviewing the options with DOT officials, learned that the main federal source of data, HPMS, could not reliably be used to analyze freight truck trends from 2007 to 2012 because, among other things, the data do not sufficiently distinguish among classes of trucks. As a result, to provide general information on trends in truck flows from 2007 to 2012, we reported information available in studies by the ATA and the American Transportation Research Institute, the research arm of ATA, which performs some of its studies in partnership with the Federal Highway Administration (FHWA). We reviewed DOT's evaluation of the ATA truck tonnage data and the methodologies used in the American Transportation Research Institute studies and determined that they were sufficiently reliable to use to provide contextual information. We also obtained some information on freight truck levels in 2013 from ATA and ATRI and interviewed ATRI, ATA, Owner-Operator Independent Drivers Association, and DOT officials.

To present a map of public highway-rail grade crossings that had over certain thresholds of both rail and vehicular traffic, we analyzed data from three sources: FRA's National Grade Crossing Inventory for the location of highway-rail grade crossings, DOT's 2012 HPMS data on average annual daily traffic (AADT) for vehicular traffic, and STB's 2012 data on rail carloads annually, as flowed over the rail lines by FRA. We used DOT's 2012 HPMS data and STB's 2012 rail data after determining that it was the most recent, reliable national data available for each type of traffic. Through discussions with DOT officials, we determined that HPMS's vehicular count data as measured by AADT (as opposed to its truck count data discussed above) are sufficiently reliable for purposes of our review. Our analysis showed that 44,982 grade crossings identified in FRA's National Grade Crossing Inventory were on roads for which 2012 HPMS data included AADT information (all federal-aid highways and highways designated on the National Highway System, covering about 25 percent of all public roads) and on rail lines that were included in FRA's flow of STB rail data for 2012 (rail lines used by rail carriers terminating 4,500 or more revenue carloads annually). Because of the lack of data on congestion at rail crossings, we were unable to identify the thresholds for levels of vehicular traffic or of train carload traffic above which a highway-rail grade crossing becomes problematic to a community due to local traffic congestion impacts. Moreover, from our interviews and case studies, we had determined that there are a number of factors that feed into how problematic a community considers a highway-rail grade crossing, such as the centrality of the crossing to a community and the extent to which emergency vehicles are located on one side of the crossing. Furthermore, our data selection, as described, only includes certain highways and railways, and there may be other roads and railways where communities experience local traffic congestion impacts of rail traffic. Nonetheless, we determined that it would be useful to display the geographic distribution of crossings that exceed these thresholds of vehicular and rail carload traffic.

After considering the thresholds used to by FRA to identify ranges of carload numbers presented in national maps of rail flows, and after considering AADT volume group ranges as identified by DOT in its HPMS field guide, we selected the threshold of 200,000 or more rail carloads annually and 10,000 or more AADT. We selected these thresholds because they represented crossings with higher vehicular and train carload traffic levels and because higher traffic levels are related to congestion. Because we lacked data that would directly link these thresholds to congestion, however, it is possible that some crossings above these thresholds do not experience congestion while other crossings below these thresholds do experience congestion. Selecting lower thresholds would produce more crossings and selecting higher thresholds would produce fewer crossings and would result in different spatial patterns. We identified that 3,762 crossings met both of these thresholds. These crossings represent about 8 percent of all crossings in our population. We reviewed the reliability of these data by reviewing documentation from the agency on the collection of this data and interviewing agency officials about their data collection procedures. We determined that these data were sufficiently reliable for the purposes they were used for in this report.

To determine how freight-related traffic congestion impacts communities, how communities have funded mitigation efforts for freight-related traffic congestion, and what challenges, if any, exist, we selected twelve locations in seven states: Chicago, Illinois; Seattle, Spokane, and Edmonds, Washington; Billings and Glendive, Montana; Williston (and surrounding communities in the Bakken oil field region), North Dakota; Gulfport and Pascagoula, Mississippi; Miami, Florida; Paulsboro, New Jersey and the New York City metropolitan region, including projects for New Jersey and New York. We selected these locations to provide a range of population sizes and geographic locations. Sizes include a mix of three large cities (population greater than 500,000); three medium cities (population between 100,000 and 500,000); and six small, communities (population less than 100,000). The locations are in the east/mid Atlantic, south, mid-west, west and northwest regions of the country. We also selected locations with differing types of federal funding obtained, such as Transportation Infrastructure Finance and Innovation Act (TIFIA) assistance or Transportation Investment Generating Economic Recovery (TIGER) grants, and locations with and without marine ports.

Where feasible, at each location, we interviewed state and local officials, railroads, metropolitan planning organizations (MPO), port officials, and private contractors, and toured congested areas with mitigation activities, such as improved intermodal connections, the relocation of rail interchanges, and grade separation projects. Where available we obtained information on the costs of mitigation projects and studies documenting and measuring freight congestion impacts. We reviewed the methodology and assumptions of these studies. We also reviewed information on transportation trends and conditions and state and local impacts of transporting certain commodities such as coal and oil, as well as the local impacts from rail and highway freight-transportation operations including plans by localities for new truck routes to bypass their downtown. Where available we obtained state level documents and plans including state freight and rail plans. The results of the interviews at these locations are not generalizable to all locations experiencing freight-related traffic congestion. However, they do provide the perspectives and experiences of various types of communities on local congestion impacts related to freight and on efforts to mitigate these impacts.

To understand the extent that MAP-21 addresses freight-related traffic congestion in communities, we reviewed federal laws and programs, including MAP-21 freight provisions, and documentation of DOT's efforts to implement MAP-21's freight provisions, such as its draft primary freight network, as well as DOT's April 2014 surface transportation reauthorization proposal. We also reviewed published draft recommendations of the National Freight Advisory Committee and attended two public meetings of this committee, and we interviewed DOT officials about ongoing efforts to implement MAP-21's freight provisions.

To inform our understanding of all of these issues, we also analyzed documents related to these issues, including legislation and prior GAO work on surface transportation and capital planning issues. In addition, we interviewed numerous freight transportation stakeholders. This included officials from federal agencies, including DOT and its modal administrations FRA, FHWA—including officials in its national and division offices—and the Bureau of Transportation Statistics, as well as STB. As part of our case study work, we interviewed or obtained prepared responses from state and local organizations including the Washington Department of Transportation; Seattle Department of Transportation and local officials in Edmonds, Washington; the Illinois Department of Transportation and Illinois Commerce Commission; Chicago Department of Transportation officials as part of the CREATE Program, the North Dakota Department of Transportation and local elected officials in Williston, ND, and the Three-Affiliated Tribal government in New Town, ND; and the City of Billings, MT. We also interviewed the transportation departments in New Jersey, Florida, Mississippi, and Montana including the Yellowstone County Commission and Planning Board in Billings, MT. We also interviewed a non-profit agency and representatives of academia and a specific rail terminal project in Bellingham, WA. We interviewed the Association of Metropolitan Planning Organizations and MPOs located in Spokane and Bellingham, Washington; The Delaware Valley Region of Pennsylvania and New Jersey; Biloxi, Mississippi; and Miami-Dade, Florida.

We also interviewed or received written responses from industry stakeholders representing different freight transportation modes, including select U.S. railroad companies—Burlington Northern Santa Fe, the Union Pacific, CSX, and Norfolk Southern Corporation—and additional railroad partners involved with the CREATE Program including Belt Railway Company railroads in the Chicago Transportation and Coordination Office. We interviewed officials from industry associations including the American Association of Railroads; the American Trucking Association; Intermodal Association of North America; Owner-Operator Independent Drivers Association; the American Association of State Transportation Officials; and the World Shipping Council.

We conducted this performance audit from September 2013 to September 2014 in accordance with generally accepted government auditing standards. Those standards require that we plan and perform the audit to obtain sufficient, appropriate evidence to provide a reasonable basis for our findings and conclusions based on our audit objectives. We believe that the evidence obtained provides a reasonable basis for our findings and conclusions based on our audit objectives.

End Notes

[1] GAO. Freight Transportation: National Policy and Strategies Can Help Improve Freight Mobility. GAO-08-287(Washington, D.C., Jan. 7, 2008).

[2] Freight-related traffic congestion can also affect the environment, such as by causing air pollution, and safety. We have recently issued a report focused on safety related to rail. See GAO. Rail Safety: Improved Capital Planning Could Address Emerging Safety Oversight Challenges. GAO-14-85 (Washington, D.C., Dec. 9, 2013).

[3] This report focuses on freight-related traffic congestion caused by rail and truck freight movements because rail and truck shipments comprise the majority of freight movements by weight. In addition, a survey by the Transportation Research Board found that the most often cited issues related to freight facilities and operations were concerns about impacts on traffic flow and congestion related to highway-rail grade crossings and trucks. See Transportation Research Board of the National Academies: Integrating Freight Facilities and Operations with Community Goals: A Synthesis of Highway Practice; Washington, D.C., 2003.

[4] This request was originally made by the former Chair of the Senate Subcommittee on Transportation and Infrastructure of the Senate Committee on Environment and Public Works, Senator Max Baucus, who resigned.

[5] While Figure 1 shows tonnage of freight moved by rail in order for rail volumes to be easily comparable to truck and inland waterway volumes, the rest of this report focuses on rail carloads, since rail carloads are more closely associated with potential traffic congestion impacts in communities at highway-rail grade crossings.

[6] Oil contained in underground shale formations was previously considered to be inaccessible because traditional techniques did not yield sufficient amounts for economically viable production. The application of horizontal drilling techniques and hydraulic fracturing—a process that injects a combination of water, sand, and chemical additives under high pressure to create and maintain fractures in underground rock formations that allow oil to flow—have played a major role in the recent increases of U.S. crude oil production.

[7] ITS technologies consist of a range of communications, electronics, and computer technologies, such as systems that collect real-time traffic data and transmit information to the public via dynamic message signs and other means.

[8] According to DOT officials, another type of project to alleviate freight-related congestion is a re-design of railroad junction points that can reduce the amount of backing movements by trains that can block grade crossings for extended periods of time. DOT stated that it has reviewed projects of this type, for example, in Willmar and Moorhead, Minnesota.

[9] Traditionally, federal surface transportation's funding has been primarily delivered through formula grant programs based on distributions prescribed by federal statute. Unlike formula grant programs, discretionary grant programs are generally based on a competitive selection process in which agencies may also need to comply with specific statutory or regulatory requirements as well as published selection criteria established for a program.

[10] Some public port authorities are self-funding public entities, meaning that they rely exclusively on revenues generated from facility operations, such as tolls, user fees, and tenant leases.

[11] GAO, Highway Public Private Partnerships: More Rigorous Up-front Analysis Could Better Secure Potential Benefits and Protect the Public Interest, GAO-08-44 (Washington, D.C.: Feb. 8, 2008).

[12] GAO-08-44.

[13] Pub. L. No. 112-141, 126 Stat. 405 (2012)

[14]The following are the goals for the national freight policy: to (1) invest in infrastructure improvements and implement operational improvements that strengthen the contribution of the national freight network to the economic competitiveness of the United States; reduce congestion; and increase productivity, particularly for domestic industries and businesses that create high-value jobs; (2) improve the safety, security, and resilience of freight transportation; (3) improve the state of good repair of the national freight network; (4) use advanced technology to improve the safety and efficiency of the national freight network; (5) incorporate concepts of performance, innovation, competition, and accountability into the operation and maintenance of the national freight network; (6) improve the economic efficiency of the national freight network; and (7) to reduce the environmental impacts of freight movement on the national freight network. 23 U,S,C, § 167(b).

[15] The Safe, Accountable, Flexible, Efficient Transportation Equity Act: A Legacy for Users, (SAFETEA-LU), Pub. L. No. 109-59, 119 Stat.1144 (2005).

[16] Comments on the primary freight network were due on February 15, 2014.

[17] GAO Action Tracker on Economic Development: Surface Freight Transportation, Action 3.

[18] GAO Action Tracker on Economic Development: Surface Freight Transportation, Action 2.

[19] H.R. 4834, 113th Cong. § 1101 (2014).

[20] Due to a lag in data reporting, 2013 STB data were not available for this report. We chose to analyze STB data beginning in 2007 after being advised by STB officials that to examine recent trends, 2007 was more indicative than 2008 of rail traffic before the recent U.S. economic recession.

[21]According to AAR officials, crude oil represented an estimated 1.4 percent of total class1-originated carloads in 2013.

[22] Freight railroads also carry many other types of commodities, such as farm products, forest products, chemicals, food, textiles, and machinery. In this report, we are focusing on trends related to intermodal, coal, and crude oil transportation by rail because they were emphasized by STB and AAR officials as having a significant impact on changes in rail traffic in recent years. In addition, intermodal and coal carloads are the only two product types to make up at least 20 percent of all rail carloads in 2012.

[23] According to STB data, a seventh state, Rhode Island, also experienced an increase of at least 5 percent in the total number of rail carloads, but due to the small number of carloads involved, this increase was not statistically significant.

[24] We used STB's data's state-through indicator as the measure of whether a carload was in a state. This indicator is present when a carload originates or terminates in a state. In addition, this indicator is present when according to a model, the carload would have passed through a state given its origin and termination points.

[25] In 2005, SAFETEA-LU authorized $197.5 billion for the federal-aid highway program for fiscal years 2005 through 2009. The American Recovery and Reinvestment Act of 2009 appropriated $48.1 billion to DOT to invest in transportation infrastructure. Pub. L. No. 111-5, 123 Sat 115 (2009).

[26]Another 12 percent of crude-oil rail carloads that originated in the U.S. originated in Texas. No other state originated more than 3 percent of these carloads in 2012.

[27] GAO, Petroleum Refining: Industry's Outlook Depends on Market Changes and Key Environmental Regulations, GAO-14-249 (Washington, D.C.: Mar. 14, 2014).

[28] Due to a lag in data reporting, 2013 STB data were not available for this report. We included 2013 AAR-reported data for illustrative purposes and to provide information as reported by the industry on trends after 2012. Because the data were not used to support our findings, conclusions, or recommendations, we did not independently analyze the AAR data.

[29] The Energy Information Administration projects that total U.S. crude oil production will reach 9.6 million barrels per day in 2019, which is 3.1 million barrels per day more than in 2012.

[30] AAR officials noted that even if the amount of crude oil shipped by rail doubled, it would still be less than 3 percent of total rail traffic.

[31] Public rail crossings are those on highways under the jurisdiction of a public authority open to the traveling public. Private crossings are those on roadways privately owned and utilized only by the landowner or licensee.

[32] Vehicle counts were measured as annual average daily traffic.

[33] In December 2013, we reported that in 2012, there were 271 fatalities at highway-rail grade crossings in the U.S., and 554 trespasser fatalities. We did not focus on safety issues related to highway-rail grade crossings in this report, in part because we recently considered this issue in another report. In December 2013, we found that in addition to Federal Railroad Administration reportable train accidents, highway-rail grade crossing accidents and trespasser incidents constitute a majority of all fatalities associated with the railroad industry. See GAO-14-85.

[34] CREATE is a partnership between the State of Illinois, City of Chicago, Metra, Amtrak, the Association of American Railroads, and U.S. DOT. It is comprised of 70 rail and highway infrastructure projects in Northeast Illinois, a region that handles a quarter of the nation's freight rail volume and experiences some of the highest freight congestion levels in the country.

[35] Cambridge Systematics, Inc. for Miami-Dade Metropolitan Planning Organization, Transportation and Economic Impacts of the Freight Industry in Miami-Dade County, (December 2011).

[36] DOT's Highway Performance Monitoring System is a national-level highway-information system that includes data on the extent, condition, performance, use, and operating characteristics of the nation's highways.

[37] FRA's rail crossing inventory includes information on each rail crossing in the U.S., including the average number of daily train crossings.

[38] ATA and ATRI undertake research on freight trucking issues. ATRI performs analyses of truck congestion in partnership with FHWA, analyses that include tracking about 500,000 trucks with GPS devices.

[39] ATA officials stated that truck congestion, which ATRI studies nationally, is a good proxy for where truck flow volumes are highest.

[40] National Cooperative Highway Research Program, Report 586: Rail Freight Solutions to Roadway Congestion—Final Report and Guidebook (Washington, D.C.: Transportation Research Board, 2007).

[41] Cambridge Systematics: Transportation and Economic Impacts of the Freight Industry in Miami-Dade County (December 2011).

[42] U.S. DOT Maritime Administration, Panama Canal Expansion Study: Phase 1 Report: Developments in Trade and National and Global Economies (Nov. 2013); and U.S. Army Corps of Engineers, U.S. Port and Inland Waterways Modernization: Preparing for Post-Panamax Vessels (June 20, 2012).

[43] See appendix I for more information about our selection methodology for site visit locations and congestion mitigation projects.

[44] Local funds include those provided by public port authorities, which are often self-financing entities with their own revenue streams.

[45] Two projects—the Port of Pascagoula's Intermodal Improvements and Seattle's Duwamish ITS Project—included contributions from one of the port's terminal operators and a private foundation, respectively.

[46] This project restored and upgraded rail service between the Port of Miami and the Florida East Coast Rail Yard in Hialeah, FL, which was suspended in 2005 following damage to the rail bridge during Hurricane Wilma, and provided the Port of Miami with direct cargo access to the national rail system.

[47] 23 U.S.C. § 602(a)(9).

[48] Unlike railroads, other private entities, such as trucking firms and distribution center operators, do not own freight infrastructure. As a result, they may be less inclined to make direct contributions to specific congestion mitigation projects. For instance, according to ATA representatives, the trucking industry does not see it as its role to directly fund road infrastructure projects because this infrastructure is publicly-owned; rather, the trucking industry indirectly contributes to infrastructure projects through federal fuel taxes and other taxes, tolls, and registration fees. However, we have previously reported that at the national level, infrastructure costs attributable to commercial freight transported by trucks exceed the revenue that these freight transportation providers pay governments to fund that infrastructure. See GAO-11-134.

[49] 23 U.S.C. § 130(b), 23 C.F.R.§ 646.210(b)(3). This contribution requirement only applies to the elimination of grade crossings that have active warning devices in place or ordered to be installed by a state regulatory agency.

[50] Railroads companies determine their contributions to grade crossing elimination projects funded outside of the Railway-Highway Crossings Program; therefore these contributions may be less than or greater than 5 percent.

[51] 49 U,S.C. § 11324.

[52] STB's environmental reviews look at the direct, indirect, and cumulative impacts of the proposed construction or acquisition. They do not include pre-existing conditions, such as rail traffic levels prior to the proposed transaction. For example, if a rail line currently accommodates 15 trains a day—and due to an acquisition is expected to accommodate an additional 5 trains a day—the environmental review would only focus on the impacts of the 5 additional trains.

[53] The decision was upheld in Village of Barrington v. Surface Transportation Board 636 F. 3d 650, 394 U.S. App. DC 353 (2011).

[54] STB can address transportation emergencies by directing traffic over particular routings, but such emergency service orders are statutorily limited in duration. STB can issue orders that affect routing on a permanent basis to address competitive issues, but not simply to mitigate freight traffic congestion.

[55] For example, if traffic congestion at one grade crossing results in vehicle backups onto another grade crossing, the congestion issue at the first crossing could be addressed due to the safety issues it presents at the second crossing.

[56] 23 U.S.C. § 130(e)(1).

[57] Federal grants are typically intended to supplement the efforts of state and local governments rather than sup-plant them. To that end, many grant programs include matching requirements or maintenance-of-effort provisions that require state and local governments to partially pay for a program from nonfederal revenues.

[58] GAO, Freight Transportation: Strategies Needed to Address Planning and Financing Limitations, GAO-04-165 (Washington, D.C.: Dec. 19, 2003). GAO, Freight Transportation: National Policy and Strategies Can Help Improve Freight Mobility, GAO-

08-287 (Washington, D.C.: Jan. 7, 2008). In GAO-08-287, we recommended that DOT work with the Congress and freight stakeholders to develop a national strategy to transform the federal government's involvement in freight transportation projects. DOT's efforts to develop a national freight transportation strategy are discussed in the next section.

[59] GAO, Freight Transportation: Strategies Needed to Address Planning and Financing Limitations, GAO-04-165 (Washington, D.C.: Dec. 19, 2003).

[60] DOT is early in the implementation process, so the extent to which community impact issues will be incorporated into final results cannot yet be determined.

[61] 23 U.S.C. § 167(d)(1)(ii). According to DOT officials, the "centerline" requirement means that the length of each road segment included in the network must be counted against the mileage limit of the network. 23 U.S.C.§ 167(d)(2) provides for the designation of not more than 3,000 additional centerline mile of roadways critical to efficient movement of goods on the primary freight network.

[62] MAP-21 requires that the national freight network is to include a primary freight network, portions of the Interstate System not designated as part of the primary freight network, and critical rural freight corridors. DOT officials stated that they have waited to provide guidance on establishing the critical rural freight corridors until the primary freight network is established.

[63] As defined in MAP-21, an aerotropolis transportation system is a multimodal freight and passenger transportation network for a defined region of economic significance centered around a major airport.

[64] MAP-21 includes the following factors for DOT to consider in designating the primary freight network: (1) the origins and destinations of freight movement in the U.S.; (2) the total freight tonnage and value of freight moved by highways; (3) the percentage of annual average daily truck traffic in the annual average daily traffic on principal arterials; (4) the annual average daily truck traffic on principal arterials; (5) land and maritime ports of entry; (6) access to energy exploration, development, installation, or production areas; (7) population centers; and (8) network connectivity. 23 U.S.C. § 167(d)(1)(B).

[65] According to this analysis, of the 300 comments uploaded to the electronic document on the draft primary freight network by March 10, 2014, 168 were policy related.

[66] GAO. Executive Guide: Leading Practices in Capital Decision-Making, GAO/AIMD-99-32 (Washington, D.C., December 1998). This executive guide is based on extensive research to identify leading practices in capital decision-making used by state and local governments and private sector organizations and identifies organizational attributes that are important to the capital decision-making process as a whole, as well as capital decision-making principles and practices used by outstanding state and local governments and private sector organizations.

[67] GAO. Surface Transportation: Principles Can Guide Efforts to Restructure and Fund Federal Programs, GAO-08-744T (Washington, D.C., July 10, 2008). We identified a framework of principles that can help inform Congress in assessing federal surface transportation programs through our prior analysis of surface transportation programs as well as a body of work that we have developed for Congress, including GAO's High-Risk, Performance and Accountability, and 21st Century Challenges reports.

[68] GAO. High Speed Passenger Rail: Future Development Will Depend on Addressing Financial and Other Challenges and Establishing a Clear Federal Role. GAO-09-317 (Washington, D.C.: Mar. 19, 2009).

[69] H.R.4834, 113th Cong. § 1101 (2014).

[70] 23 U.S.C. § 167(f)(1)(H).

[71] Department of Transportation: "Interim Guidance on State Freight Plans and State Freight Advisory Committees," Federal Register, Vol. 77 No. 199 (Oct. 15, 2012). According to DOT officials, as of April 2014, about 25 states had completed a state freight plan and about 42 had either completed or stated to DOT that they planned to complete a state freight plan. About 8 states stated they are not planning to complete one. These states generally told DOT they are not completing one due to the financial burden of completing one or because freight information included in other required plans is sufficient.

[72] MAP-21, section 1116, provides that if a project is identified in an approved state freight plan, DOT can increase the federal share payable up to 95 percent for projects on the Interstate, and up to 90 percent for any other project. According to DOT officials, two states, Indiana and Vermont, have done this so far. In these cases only, DOT reviews the state freight plans. It approved Indiana's, and at the time of our discussion, Vermont's was under review.

[73] On July 14, 2014, DOT stated that the National Freight Advisory Committee's recommendations were under review by DOT.

[74] GAO. Surface Transportation: Restructured Federal Approach Needed for More Focused, Performance-Based, and Sustainable Programs, GAO-08-400 (Washington, D.C.; Mar. 6, 2008).

[75] Interim Guidance on State Freight Plans and State Advisory Committee, 77 Fed. Reg. 62596 (Oct. 15, 2012). According to DOT officials, the interim guidance on state freight plans was prepared very quickly because states were eager to know what kind of state freight plan they needed to prepare in order to satisfy the requirements of section 1116 of MAP-21, which allows states, with the approval of the Secretary, to contribute a smaller state match on certain freight-related highway projects. According to DOT, due to this short time period, it was not practical to include detailed guidance on how to document each of the elements that the statute required state freight plans to include.

[76] The other five areas are (1) Reducing congestion; (2) Improving safety and resilience; (3) Improving the state of good repair; (4) Enhancing economic efficiency, productivity, and competitiveness; and (5) Use of innovative technology, competition, performance management, and accountability.

[77] In its recent recommendations to DOT, the National Freight Advisory Committee suggested that a potentially useful federal role in this area could be to develop a repository of best practice efforts at the local level related to mitigating the negative community impacts of freight, and it recommended that the National Freight Strategic Plan should develop a set of criteria for defining best practices to be shared with freight stakeholders through the establishment of a clearinghouse of freight best practices and a program for disseminating best practices. On July 14, 2014, DOT stated that the National Freight Advisory Committee's recommendations were under review by DOT.

[78] In the Notice publishing its draft primary freight network, DOT listed multiple data sources used, including, among others, DOT's Freight Analysis Framework and Highway Performance Monitoring System.

[79] On July 14, 2014, DOT stated that the National Freight Advisory Committee's recommendations were under review by DOT.

[80] GAO/AIMD-99-32. This executive guide is based on extensive research to identify leading practices in capital decision-making used by state and local governments and private sector organizations and identifies organizational attributes that are important to the capital decision-making process as a whole, as well as capital decision-making principles and practices used by outstanding state and local governments and private sector organizations.

[81] GAO. Executive Guide: Effectively Implementing the Government Performance and Results Act. GAO/GGD-96-118 (Washington, D.C., June 1996). For this guide, at the request of Congress, we studied a number of leading public sector organizations that were successfully pursuing management reform initiatives and becoming more results-oriented, and identified key steps commonly taken by these organizations to become more results-oriented.

[82] GAO. Executive Guide: Effectively Implementing the Government Performance and Results Act. GAO/GGD-96-118 (Washington, D.C., June 1996). For this guide, at the request of Congress, we studied a number of leading public sector organizations that were successfully pursuing management reform initiatives and becoming more results-oriented, and identified key steps commonly taken by these organizations to become more results-oriented.

[83] GAO-08-744T and also see GAO. Surface Transportation: Restructured Federal Approach Needed for More Focused, Performance-Based, and Sustainable Programs, GAO-08-400 (Washington, D.C.; March 6, 2008).

[84] GAO-08-400.

INDEX

A

access, 20, 40, 43, 59, 66, 67, 73, 95, 96
accountability, 93, 97
agencies, vii, 1, 2, 4, 5, 14, 27, 41, 44, 46, 70, 71, 72, 77, 82, 90, 92
American Recovery and Reinvestment Act of 2009, 23, 52, 93
American Transportation Research Institute (ATRI), 37, 40, 41, 61, 62, 88, 94
American Trucking Associations (ATA), 37, 40, 41, 61, 88, 94, 95
Association of American Railroads (AAR), 2, 3, 5, 6, 13, 27, 30, 37, 40, 41, 49, 52, 53, 55, 87, 93, 94
authority(ies), 11, 14, 31, 44, 64, 69, 92, 94
automobiles, 6
average annual daily traffic (AADT), 37, 88, 89

B

benefits, 23, 35, 46, 55, 60, 64, 67, 68, 69, 71, 83
border crossing, 5, 16
Brotherhood of Locomotive Engineers and Trainmen (BLET), 3, 5, 11, 12, 14, 15, 16, 27, 28
Bureau of Transportation Statistics (BTS), vii, 1, 2, 3, 4, 5, 7, 8, 28, 29, 30, 32, 37, 41, 90
Burlington-Northern Santa Fe (BNSF), 30, 33, 37, 41, 53, 60, 68, 80
businesses, 20, 93

C

capital projects, 45
carbon monoxide, 45
case study(ies), 89, 90
CBP, 2, 3, 4, 5, 9, 10, 11, 12, 13, 16, 20, 21, 22, 23, 24, 27, 28, 30, 31, 32
Census, 19, 20, 23, 80
certification, 16, 32
challenges, 2, 7, 15, 23, 25, 39, 64, 71, 72, 82, 83, 86, 89
Chicago, 38, 40, 52, 56, 60, 61, 65, 68, 70, 89, 91, 94
Chicago Region Environmental and Transportation Efficiency Program (CREATE), 38, 56, 60, 65, 70, 91, 94
cities, 18, 31, 39, 61, 63, 89
coal, 6, 12, 39, 49, 50, 52, 53, 54, 55, 57, 83, 86, 90, 93
collaboration, 44, 67
commercial, 95
commodity, 39, 42, 52, 55, 61, 87
communication, 17

communities, vii, 1, 2, 4, 5, 6, 7, 9, 11, 12, 17, 18, 19, 20, 23, 24, 25, 26, 27, 28, 29, 30, 32, 35, 36, 37, 39, 40, 41, 42, 43, 44, 47, 49, 55, 56, 60, 62, 63, 64, 67, 69, 70, 71, 72, 74, 76, 77, 78, 79, 80, 81, 82, 83, 84, 85, 86, 88, 89, 90, 92, 96, 97
competition, 93, 97
competitive grant program, 46
competitiveness, 46, 93, 97
Congestion, v, 29, 35, 38, 43, 45, 49, 52, 56, 62, 63, 64, 65, 67, 69, 94
Congestion Mitigation and Air Quality Improvement (CMAQ), 38, 45, 52, 64
Congress, 35, 36, 39, 46, 48, 74, 75, 76, 77, 79, 84, 96, 98
construction, 23, 28, 29, 44, 45, 47, 65, 66, 67, 68, 69, 95
consumer goods, 6, 39
containers, 6, 13, 39, 41, 62
cooperation, 10, 26
coordination, 15, 24
cost, 23, 32, 36, 39, 48, 60, 62, 64, 66, 67, 68, 69, 70
cost-benefit analysis, 68
crude oil, 36, 39, 42, 49, 50, 52, 53, 54, 55, 57, 70, 86, 92, 93, 94
Customs and Border Protection (CBP), 2, 3, 4, 5, 9, 10, 11, 12, 13, 16, 20, 21, 22, 23, 24, 27, 28, 30, 31, 32

D

data collection, 28, 40, 79, 87, 89
data set, 79, 81
database, 9, 44, 61
decision-making process, 96, 97
Department of Homeland Security, 4
Department of Transportation, vii, 2, 3, 4, 27, 36, 38, 42, 53, 60, 63, 87, 91, 97
Department of Transportation (DOT), v, vii, 1, 2, 3, 4, 5, 9, 10, 13, 15, 16, 17, 23, 24, 25, 26, 27, 28, 29, 31, 32, 35, 36, 37, 38, 39, 40, 41, 42, 44, 45, 46, 47, 48, 52, 53, 56, 58, 60, 61, 62, 63, 66, 68, 70, 71, 72, 73, 74, 75, 76, 77, 78, 79, 80, 81, 82, 83, 85, 86, 87, 88, 89, 90, 92, 93, 94, 96, 97
distribution, 43, 44, 89, 95
draft, 26, 27, 37, 41, 48, 73, 74, 75, 79, 84, 85, 90, 96, 97

E

economic competitiveness, 46, 93
economic efficiency, 93, 97
economic growth, 39
emergency, 7, 42, 59, 81, 89, 95
emergency response, 42
employees, 31, 32
energy, 39, 55, 62, 67, 73, 96
enforcement, 30
environment, 68, 92
environmental impact, 47, 69, 72, 77, 93
environmental sustainability, 46
equipment, 4, 9, 12, 14, 44
evidence, 6, 41, 91
exports, 53, 55
extraction, 49, 55, 62

F

federal aid, 80
federal funds, 52, 64
federal government, 47, 52, 79, 96
Federal Highway Administration, 38, 40, 41, 45, 88
Federal Highway Administration (FHWA), 38, 40, 41, 44, 45, 61, 68, 88, 90, 94
federal law, 36, 41, 90
Federal Railroad Administration (FRA), 2, 3, 4, 5, 7, 9, 14, 15, 16, 19, 20, 21, 22, 24, 25, 27, 28, 31, 33, 38, 40, 41, 44, 46, 60, 61, 81, 87, 88, 89, 90, 94
Federal Register, 97
financial, 46, 68, 97
fiscal year, 46, 93

Fleming, Susan A., 27, 86
flexibility, 55, 74
formula, 44, 64, 92
freight flows, vii, 4, 35, 36, 39, 40, 47, 49, 56, 72, 77, 80, 86, 87, 88
freight rail, vii, 1, 2, 3, 4, 5, 7, 9, 10, 12, 14, 15, 17, 20, 23, 24, 25, 27, 28, 30, 33, 36, 46, 49, 53, 69, 94
freight rail operations, vii, 2, 5, 27, 28
funding, 7, 21, 23, 25, 26, 39, 44, 46, 48, 63, 64, 65, 66, 68, 69, 70, 71, 72, 75, 82, 83, 86, 90, 92
funds, 46, 52, 64, 66, 70, 71, 94

G

GAO, 1, 2, 8, 10, 13, 17, 18, 19, 21, 22, 27, 29, 30, 32, 33, 35, 36, 41, 42, 43, 46, 50, 51, 54, 57, 58, 59, 63, 66, 90, 92, 93, 94, 95, 96, 97, 98
General Motors, 30
Government Performance and Results Act (GPRA), 38, 82, 98
governments, 2, 3, 7, 25, 26, 46, 63, 75, 77, 95, 96, 97
grant programs, 44, 92, 95
grants, 65, 67, 90, 95
growth, 7, 30, 39, 52, 55
guidance, 26, 60, 77, 78, 81, 82, 83, 84, 96, 97
Gulf Coast, 52, 53, 62

H

harmonization, 15, 16, 17, 31
high-risk shipments, 9
Highway Performance Monitoring System (HPMS), 38, 80, 81, 87, 88, 89, 94, 97
highway system, 61, 80
highway-rail grade crossings, vii, 1, 2, 4, 5, 7, 9, 11, 12, 13, 14, 17, 18, 19, 20, 21, 22, 23, 24, 25, 26, 27, 28, 32, 33, 36, 39, 42, 43, 55, 56, 58, 60, 70, 71, 74, 80, 81, 86, 88, 92, 94
highways, 38, 39, 41, 42, 43, 47, 71, 73, 74, 75, 76, 80, 87, 88, 94, 96
House, 1, 4, 5, 26, 48, 76, 86
House of Representatives, 4, 48, 76, 86
Housing and Urban Development, 1, 5

I

identification, 47, 72, 77
improvements, 9, 25, 26, 40, 45, 46, 47, 65, 66, 68, 70, 93
industry, 7, 25, 41, 52, 53, 55, 60, 75, 79, 91, 93, 94, 95
infrastructure, 5, 7, 9, 11, 12, 23, 26, 28, 44, 45, 46, 47, 53, 64, 70, 71, 93, 94, 95
inspections, 2, 4, 10, 11, 12, 13, 14, 15, 17, 20, 27, 30, 31
intelligence, 30
intelligent transportation systems (ITS), 38, 43, 65, 92, 95
international freight rail, vii, 1, 2, 4, 5, 18, 27, 29, 30
international trade, 10
investment, 9, 13, 15, 26, 44, 45, 47, 48, 67, 75, 76, 77
issues, 9, 10, 15, 21, 24, 27, 36, 40, 41, 43, 44, 56, 60, 64, 68, 69, 70, 71, 72, 76, 77, 78, 79, 81, 82, 83, 90, 92, 94, 95, 96

J

jurisdiction, 69, 94

L

laws, 16, 36, 41
lawyers, 22
leadership, 44
legislation, 39, 41, 47, 90
loan guarantees, 45

local community, 26, 84
local government, 3, 7, 25, 26, 46, 63, 75, 95, 96, 97

M

majority, 4, 22, 41, 53, 69, 92, 94
management, 45, 46, 97, 98
manufactured goods, 4
manufacturing, 13
methodology, 41, 87, 90, 94
metropolitan areas, 74, 80
metropolitan planning organization (MPO), 3, 5, 18, 22, 24, 29, 30, 32, 38, 71, 77, 90
Mexico, vii, 1, 2, 4, 6, 7, 10, 13, 14, 15, 16, 23, 24, 26, 27, 30, 31, 32
Miami, 36, 40, 60, 62, 65, 66, 67, 71, 89, 91, 94, 95
Montana, 40, 55, 56, 59, 60, 70, 71, 89, 91
movement restrictions, 31
Moving Ahead for Progress in the 21st Century Act (MAP-21), vii, 3, 9, 35, 36, 37, 38, 39, 41, 44, 47, 48, 72, 73, 76, 77, 78, 83, 84, 85, 86, 90, 96, 97

N

National Ambient Air Quality Standards, 45
National Highway System (NHS), 38, 44, 45, 73, 88
national security, 2, 10, 12
National Strategy, v, 29, 35, 48, 96
negative effects, 18, 60
New Jersey DOT (NJ DOT), 38, 70, 71
North America, 16, 30, 91

O

Obama, President Barack, 26
officials, 2, 3, 5, 9, 10, 11, 12, 13, 14, 15, 16, 17, 18, 19, 20, 21, 22, 23, 24, 25, 27, 28, 29, 30, 31, 32, 39, 40, 41, 47, 48, 49, 52, 53, 55, 56, 59, 60, 61, 62, 67, 68, 69, 70, 71, 72, 74, 75, 77, 78, 80, 81, 82, 86, 87, 88, 89, 90, 91, 92, 93, 94, 96, 97
oil, 36, 39, 40, 42, 49, 50, 52, 53, 54, 55, 57, 61, 63, 70, 74, 83, 87, 89, 90, 92, 93, 94
oil production, 36, 42, 53, 61, 63, 92, 94
omission, 75
operating revenues, 30
operations, vii, 2, 5, 12, 26, 27, 28, 44, 47, 67, 81, 90, 92

P

Pacific, 30, 33, 38, 41, 68, 91
Panama, 62, 94
participants, 67
Petroleum, 93
pipeline, 44, 53
plants, 6, 7
POEs, vii, 1, 2, 3, 4, 5, 7, 9, 11, 12, 16, 17, 25, 28, 29, 30, 31
policy, 9, 36, 41, 47, 72, 74, 76, 77, 93, 96
policymakers, 75, 83
population, 20, 30, 73, 74, 89, 96
population size, 89
port of entry (POE), vii, 2, 3, 4, 5, 9, 11, 12, 14, 18, 19, 20, 23, 24, 25, 26, 27, 28, 29, 30, 31, 32
President, 26, 39, 47
principles, 68, 75, 77, 82, 96, 97
private benefits, 68
Private Funding, 65
private sector, 46, 64, 66, 71, 76, 96, 97
project, 23, 32, 45, 47, 52, 56, 64, 66, 67, 69, 70, 71, 91, 92, 95, 97
project sponsors, 66, 71
public interest, 69
public sector, 46, 64, 67, 69, 98
public support, 71
public-private partnerships, 46, 66, 67
publishing, 97

Q

quality of life, 7, 78

R

Rail Vehicle and Cargo Inspection System (R-VACIS), 2, 3, 9, 10, 11, 12, 13, 20, 21, 24, 31
recession, 36, 40, 49, 52, 61, 86, 93
recommendations, iv, 2, 26, 28, 36, 48, 79, 90, 93, 97
reconstruction, 66
regulations, 2, 9, 15, 16, 17, 27, 31, 44, 68, 75
regulatory requirements, 92
relevance, 82, 86
reliability, 9, 29, 40, 87, 89
renewable energy, 67
repair, 46, 70, 93, 97
requirement, 6, 9, 14, 17, 24, 28, 31, 36, 67, 68, 73, 75, 76, 77, 82, 83, 84, 86, 95, 96, 97
resilience, 93, 97
resources, 47, 67, 73, 75, 76, 83
response, 6, 26, 28, 42, 55
restrictions, 9, 31
restructuring, 15
revenue, 44, 47, 55, 68, 71, 88, 94, 95
risk, 9, 10, 12, 32, 33, 46
routes, 40, 43, 90
rules, 16, 75
rural areas, 62, 79, 85

S

Safe, Accountable, Flexible, Efficient (SAFETEA-LU), 38, 52, 93
safety, 2, 4, 9, 10, 14, 15, 17, 31, 33, 44, 45, 46, 68, 70, 92, 93, 94, 95, 97
security, 2, 9, 10, 11, 12, 15, 16, 93
Senate, 4, 26, 38, 92
stakeholders, 7, 9, 11, 13, 17, 26, 36, 37, 44, 46, 47, 48, 59, 61, 70, 73, 75, 81, 85, 90, 91, 96, 97
state, 2, 3, 5, 7, 9, 11, 23, 24, 25, 26, 28, 31, 32, 33, 36, 41, 44, 46, 47, 48, 53, 55, 56, 60, 62, 63, 64, 70, 71, 72, 73, 75, 77, 78, 80, 82, 83, 84, 85, 87, 90, 91, 93, 95, 96, 97
Surface Transportation Board (STB), 38, 39, 41, 42, 44, 49, 50, 51, 54, 57, 58, 59, 68, 69, 86, 87, 88, 90, 93, 95
sustainability, 46

T

taxes, 95
technical assistance, 44
technical comments, 2, 27, 86
technology, 9, 93, 97
terminals, 39, 55
threats, 2, 9, 10, 12, 31
toll revenue, 47
tracks, 7, 36, 56
trade, vii, 1, 4, 9, 10
trading partners, 6
transport, 16, 35, 38, 59
transportation, 5, 6, 7, 9, 10, 23, 25, 27, 28, 30, 31, 37, 38, 39, 41, 43, 44, 45, 46, 47, 48, 53, 62, 64, 65, 66, 70, 71, 72, 73, 75, 76, 77, 80, 82, 85, 86, 87, 90, 91, 92, 93, 95, 96
Transportation Equity Act
A Legacy for Users, 38, 52, 93
transportation infrastructure, 46, 93
Transportation Infrastructure Finance and Innovation Act (TIFIA), 38, 45, 64, 67, 90
Transportation Investment Generating Economic Recovery (TIGER), 38, 46, 52, 60, 65, 66, 90

U

U.S. Army Corps of Engineers, 94

Union Pacific (UP), 30, 33, 38, 41, 68, 91
United States, v, vii, 1, 4, 5, 6, 7, 8, 9, 10, 11, 13, 14, 15, 16, 17, 23, 24, 28, 29, 30, 31, 32, 35, 38, 48, 53, 93
urban, 76, 79, 85

V

vehicle traffic congestion, vii, 1
vehicles, 5, 7, 15, 40, 42, 56, 59, 60, 61, 62, 63, 81, 89
violent crime, 31
vision, 16, 73, 75, 78

W

waiver, 14, 17, 27, 31
Washington, 5, 10, 20, 21, 28, 29, 38, 39, 40, 55, 56, 59, 61, 70, 71, 73, 89, 91, 92, 93, 94, 95, 96, 97, 98
Washington State DOT (WSDOT), 38, 70
waterways, 38, 41, 87
working groups, 10, 26, 27